I0475232

EasyTerms™
Terminology Guidebook
for Biochemistry

Copyright 2009, Ed Creager

This edition of EasyTerms is one in a series of simple-to-use, college-level terminology guidebooks.

Although these guidebooks were originally intended for college students, many high school students will also find them helpful as they prepare for college.

Other topics covered in existing or forthcoming editions:

- Anatomy & Physiology (Human)
- Biology
- Botany
- Business Management
- Cell Biology
- Ecology

- Genetics
- Microbiology
- Nursing
- Nutrition
- Psychology
- Zoology

EasyTerms can help support your educational advancement and can boost the vocabulary of almost anyone who reads it.

For more information on these and other publications, please visit this site:

www.ApplecreekBooks.weebly.com

and please note the author's "signature book" entitled,

"The Money-Saving Idea Book:
Inside Tips for Starving, Students, Frugal Seniors and Every Financial Survivor."

("The Money-Saving Idea Book" © and ™, Ed Creager, 2009.)

Foreword

This Biochemistry edition is a simple-to-use, college-level* terminology guidebook and is part of the EasyTerms reference series. In the book, terms are arranged alphabetically within appropriate topic areas. The complete index makes it easy to find any term and its definition.

* These books can also help high school students prepare so that, before they attend college, they'll already know a considerable amount of the terminology they'll need.

A substantial number of the terms defined here have additional definitions outside the scope of the subject being covered. More general definitions and additional meanings, if sought, are to be found in less specialized publications such as dictionaries and encyclopedias.

Please check this website...

www.ApplecreekBooks.weebly.com

...for more information on other available books.

You can sign up at the website for discount information on a variety of books, or visit now and use any of these discount codes to save 20%:

EasyTerms Human Anatomy & Physiology edition: JP2Q9KW3
EasyTerms Biology edition: P7CMFZ3B
EasyTerms Zoology edition: 3FDNAGQ5
The Money-Saving Idea Book: FPKPZY6Q

EasyTerms™
Terminology Guidebook

Table of Contents

The terms that follow are divided into the topics shown below. The page number on which the topic begins is given. Within each topic, the terms are arranged alphabetically.

Basic Chemistry...................................... 1

Introduction to Biomolecules................ 7

Cells and Organelles.............................. 13

Membranes and Movement................... 18

Carbohydrates / Lipids / Proteins......... 22

Enzymes... 27

General Energy Considerations............ 30

Photosynthesis...................................... 36

Nucleic Acid Metab./DNA Replication.. 41

Protein Synthesis.................................. 46

Digestion / Transport of Nutrients........ 52

Nutrition.. 57

Hormones... 63

Gas Exchange and Respiration............. 70

Body Fluids and Circulation................... 72

Skin and Connective Tissues................ 78

Nervous System..................................... 81

Muscle.. 86

Chemical Senses / Vision...................... 89

Body Defenses....................................... 91

Body Defense Disorders........................ 97

Acid-Base Balance / Electrolytes /
   Nitrogenous Wastes........................... 100

Basic Genetics / Mutations................... 105

Genetic Disorders / Cancer.................. 108

Biosphere... 112

Index... 116

Basic Chemistry

1. acid

An ionizing substance that donates hydrogen ions.

2. alkaline

Basic, able to accept hydrogen ions.

3. anion

A negatively charge ion.

4. atom

Smallest particle that retains properties of an element.

5. atomic number

The number of protons in the nucleus of an atom.

6. atomic symbol

One- or two-letter symbol used to stand for a particular element.

7. atomic weight

The total number of protons and neutrons in an atom; the average number if there are isotopes of the element.

8. Avogadro's number

Number of molecules in a gram molecular weight of any compound (6.023 x 10 to the 23rd power).

9. base

An ionizing substance that accepts hydrogen ions or reacts with an acid to form a salt.

10. buffer

A substance that resists pH change by holding or releasing hydrogen ions in a solution.

11. catalyst

A substance that increases a chemical reaction rate.

12. cation

A positively charged ion.

13. chemical bond

Any force produced by interacting electrons that helps to hold molecules together.

14. chemical energy

Energy stored in molecules.

15. colligative property

Property of a solution determined by the concentration of solute particles in it.

16. colloid

Glue-like; a particle in a colloidal dispersion.

17. colloidal dispersion

A state of matter with small particles suspended in a medium.

18. compound

A substance with two or more elements combined in definite proportion.

19. covalent bond

A chemical bond formed by shared electrons between two atoms.

20. Dalton

Weight of a single hydrogen atom (1.66 X 1/10 to the 24th power).

21. dipole

Molecule having both positive and negative charge.

22. diprotic acid

Acid that can donate two protons.

23. electron

A negatively charged particle that continually moves around the nucleus of an atom.

24. element

A fundamental unit of matter.

25. gram molecular weight

The quantity of a substance (in grams) equal to its molecular weight.

26. hydronium ion

Hydrated hydrogen ion.

27. ion

A charged atom or group of atoms.

28. ionic bond

A chemical bond with atoms held together by the attraction of unlike charges.

29. ionizing radiation

Radiation that causes some organic molecules to lose electrons and become more reactive.

30. isotope

An atom having a different number of neutrons than certain other atoms of the same element.

31. law of mass action

Rate of a reaction is proportional to product of activities of reactants.

32. mixture

Two or more substances combined in any proportions and retaining their individual properties.

33.   molal solution

Solution made of one mole of solute and 1000 g water.

34.   molar solution

Soluting having one mole of solute and water to make 1000 ml.

35.   mole

A gram molecular weight.

36.   molecule

The smallest quantity of a substance that retains its chemical properties.

37.   monoprotic acid

Acid capable of donating one proton.

38.   neutron

An uncharged particle in the nucleus of an atom.

39.   nucleus

Central part of an atom or a cell; a group of cell bodies in the central nervous system.

40.   organic

Containing carbon.

41.   oxidation

Addition of oxygen or loss of electrons in a chemical reaction.

42.   oxidation-reduction

Transfer of electrons from one molecule to another.

43.   oxidizing agent

Molecule that accepts electrons in an oxidation-reduction reaction.

44.   pH

A scale for expressing acidity or alkalinity; the negative logarithm of the hydrogen ion concentration.

45.   polar group

Hydrophilic group.

46.   product

A substance formed, as in a chemical reaction or process.

47.   proton

A positively charged particle in the nucleus of an atom.

48.   radiation

Spreading from a center; giving off electromagnetic particles and waves.

49.   radioactivity

Spontaneous decay of an isotope with emission of particles and energy.

50.   radioisotope

Isotope that displays radioactivity.

51.   reactant

A substance changed by a chemical reaction.

52.   reduction

Gain of an electron or loss of oxygen in a chemical reaction.

53.   salt

Compound that fully ionizes in solution.

54.   suspension

Dispersal of particles in a liquid.

55. valence

An ion's charge.

Introduction to Biomolecules

56. **absolute configuration**

Arrangement of four different substituents around asymmetric carbon atom, relative to D- and L-glyceraldehyde.

57. **alkaloid**

Nitrogenous, organic compound of plant origin, such as caffeine.

58. **amino acid**

A molecule having both acid and amino functional groups.

59. **amphipathic compound**

Substance whose molecules have polar and nonpolar regions.

60. **amphoteric compound**

Substance that can both donate and accept protons.

61. **Angstrom unit**

Length equal to one billionth of a meter used to express the size of molecules.

62. **asymmetric carbon atom**

Carbon atom covalently bonded to four substituents.

63. **biomolecule**

Organic compound normally found in a living organism.

64. **building-block molecule**

Small biomolecule that appears as a structural unit of a larger molecule.

65. **carbohydrate**

An organic compound having several alcohol groups and an aldehyde or ketone group.

66. cellulose

Fibrous carbohydrate that forms the structure of many plants and provides insoluble fiber in the human diet.

67. cholesterol

Animal steroid found in cell membranes.

68. configuration

Spatial arrangement of substituent groups about an asymmetric carbon atom.

69. conformation

The three-dimensional shape of a large molecule.

70. dehydration

Removal of water.

71. dextrorotatory isomer

Stereoisomer that rotates a plane of polarized light to the right.

72. diasteriosiomer

One of a pair of isomers about a second asymmetric carbon for each configuration at the first asymmetric carbon.

73. disaccharide

A molecule having two sugar (saccharide) units held together by a glycosidic bond.

74. enantiomer

One of a pair of mirror-image, but not superimposable, isomers.

75. epimer

One of two stereoisomers that differ in configuration at one asymmetric carbon, regardless of other asymmetric carbons.

76. fatty acid

A long hydrocarbon chain with a carboxyl group at one end.

77. functional group

A component of a molecule that participates in a chemical reaction.

78. glucose

Main blood sugar.

79. glycerol

Alcohol to which fatty acids are bound in fats.

80. glycolipid

A molecule that contains both carbohydrate and lipid components.

81. hydrogen bond

Weak covalent bond between hydrogen and another element, such as oxygen or nitrogen.

82. hydrolysis

The splitting of a molecule with the addition of water.

83. informational molecule

Molecule that has information coded in the sequence of its building blocks; protein or nucleic acid.

84. isomer

A molecule having the same kinds and number of atoms as another molecule, but arranged differently.

85. lactic acid

Product of anaerobic metabolism especially in skeletal muscle.

86. lecithin

A phospholipid characteristic of animal tissues.

87. levorotatory isomer

Isomer that rotates a plane of polarized light to the left.

88. lipid

Any fat or fatlike substance.

89. macromolecule

Molecule having a molecular weight over several thousand.

90. monosaccharide

A simple sugar.

91. native conformation

Biologically active conformation of a protein.

92. nonpolar

Uncharged; lacking polarity.

93. nonpolar group

Hydrophobic group, usually a hydrocarbon.

94. nucleic acid

A polymer of nucleotides; DNA or RNA.

95. optical activity

Ability to rotate a plane of polarized light.

96. phospholipid

A lipid made of glycerol, fatty acids, and phosphoric acid.

97. polar compound

A molecule having a charged area or polarity.

98. polymer

A molecule consisting of repeating units.

99. polysaccharide

A molecule consisting of many saccharide units connected by glycosidic bonds.

100. protein

A polymer of amino acids.

101. saturated fatty acid

A fatty acid lacking double bonds in the carbon chain and being saturated with hydrogen.

102. saturation

Condition of having all chemical affinities satisfied.

103. secretion

A cell product; the act of transporting a substance across a membrane.

104. specificity

The attribute of being specific.

105. stereoisomer

Compound having the same kind and number of atoms as another compound, but in a different spatial arrangement.

106. steroid

A lipid with a complex four-ring structure.

107. template

Pattern.

108. trace element

A chemical element normally present in very small amounts in the body.

109. triacylglycerol

A lipid molecule containing glycerol and three fatty acids.

110. **triglyceride**

A triacylglycerol (glycerol and three fatty acids).

111. **unsaturated fatty acid**

Fatty acid with pairs of hydrogen atoms replaced by double bonds in the carbon chain.

Cells and Organelles

112. **cell**

A basic functional unit of a living organism.

113. **cell membrane**

Lipid and protein compounds that form the boundary of a cell.

114. **cell theory**

A theory stating that living things are composed of cells.

115. **centriole**

One of a pair of intracellular bodies that participate in forming a mitotic spindle.

116. **chromatin**

Nuclear material that condenses into distinct chromosomes during cell division.

117. **chromosome**

In a human cell, one of 46 nuclear structures made of DNA and protein.

118. **cilium**

A tiny hairlike projection found on some epithelial cells.

119. **cisterna**

Reservoir or cavity.

120. **cytoplasm**

Cell substance, excluding the nucleus.

121. **cytoskeleton**

The organelles forming a cell's internal framework.

122. **cytosol**

The fluid part of cytoplasm that suspends organelles.

123. **endoplasmic reticulum**

A membranous vesicular network within a cell.

124. **eukaryote**

Cell having a membrane-bound nucleus and organelles.

125. **flagellum**

A movable hairlike process on a cell.

126. **gel**

A semi-solid state of a colloidal dispersion.

127. **Golgi apparatus**

Membranous vesicles clustered in cells that complete synthesis of secretions.

128. **hydrophilic**

Attacted to water.

129. **hydrophobic**

Tending to avoid water.

130. **hydrophobic interaction**

Association of nonpolar groups away from water in aqueous systems.

131. **in vitro**

In glass.

132. **in vivo**

In life.

133. **intracellular**

Within a cell.

134. **lysosome**

Membrane-bound organelle that contains digestive enzymes.

135. **metabolic turnover**

Constant degradation and replacement of cell components.

136. **microfilament**

A small, hollow protein fiber in cytoplasm that aids in movement or forms part of a cytoskeleton.

137. **microsome**

Membrane-bound vesicle derived from fragmentation of the endoplasmic reticulum.

138. **microtubule**

A cylindrical organelle that forms part of a cell's mitotic spindle.

139. **mitochondrion**

An organelle that contains enzymes for oxidative and energy-capturing processes.

140. **nuclear**

Of the nucleus.

141. **nucleolus**

A body containing RNA within a nucleus.

142. **nucleoplasm**

The substance of a nucleus.

143. **organelle**

A functional unit inside a cell.

144. osmosis

Diffusion of water from its own region of higher to lower concentration.

145. peroxisome

An organelle that contains catalase and other oxidative enzymes.

146. pinocytosis

Intake of extracellular fluid by cells; cell drinking.

147. plastid

Self-replicating plant organelle that can differentiate into a chloroplast.

148. prokaryote

Unicellular organism with circular chromosome lacking membrane-bound organelles and nucleus.

149. protoplasm

Cell substance; literally, first formed.

150. ribosomal RNA (rRNA)

A nucleic acid that forms part of a ribosome.

151. ribosome

An organelle containing ribonucleic acid and protein where protein synthesis occurs.

152. sodium-potassium pump

Mechanism that actively moves Na ions out of cells and K ions into them against gradients.

153. sol

A liquid state of a colloidal dispersion.

154. surface-to-volume ratio

The surface area of a structure divided by its volume.

155.   **tubulin**

A protein that forms intracellular microtubules.

156.   **tumor**

Abnormal aggregation of cells that can be malignant or benign.

157.   **vesicle**

A small sac filled with liquid.

Membranes and Movement

158. **active transport**

Transport of a substance against a gradient using a carrier molecule, enzyme, and cellular energy.

159. **adsorptive endocytosis**

Entry of a substance into a cell by first attaching to the cell membrane.

160. **binding site**

A site where a particular molecule binds to a membrane or other structure.

161. **bulk flow**

Streaming of molecules that allows them to move faster than by diffusion.

162. **carrier**

A transfer molecule; a person capable of transmitting an unexpressed gene.

163. **carrier saturation**

A condition with all carrier molecules carrying a substance.

164. **collagen**

A fibrous protein in connective tissue.

165. **colloidal osmotic pressure**

Pressure exerted on a membrane by colloidal particles, such as blood proteins.

166. **concentration gradient**

Range of concentrations of a substance from one region to another.

167. **diffusion**

Movement of molecules or ions from a region of higher to a region of lower concentration.

168.   **electrochemical gradient**

Total gradient from concentration and electrical charge.

169.   **endocytosis**

Movement of particles across a membrane into a cell.

170.   **exocytosis**

The movement of particles across a membrane out of a cell.

171.   **extracellular**

Outside a cell.

172.   **facilitated diffusion**

Diffusion down a gradient on carrier molecule but not requiring cellular energy.

173.   **fibrocyte**

Mature fibroblast that maintains the fibrous matrix of a connective tissue.

174.   **filtration**

Passage of a fluid across a membrane by mechanical pressure.

175.   **fluid pinocytosis**

Movement of small quantities of fluid across a membrane into a cell; cell drinking.

176.   **fluid-mosaic model**

A model of molecular arrangements in a cell membrane.

177.   **glycoprotein**

A molecule that contains both carbohydrate and protein components.

178.   **gradient**

The rate of change in the magnitude of concentration, pressure, or other variable.

179. **hydrostatic pressure**

Force exerted by a fluid.

180. **hyperosmotic**

Having higher osmotic pressure than a reference solution.

181. **hypertonic**

Causing movement of water out of cells.

182. **hyposmotic**

Having lower osmotic pressure than a reference solution.

183. **hypotonic**

Causing movement of water into cells.

184. **isosmotic**

Having the same osmotic pressure as a reference solution.

185. **isotonic**

Causing no net water movement across a cell membrane.

186. **ligand**

That which binds to a receptor.

187. **osmolarity**

A solution's osmotic concentration determined by the number of osmotically active particles it contains.

188. **osmotic pressure**

Pressure created by osmosis.

189. **passive transport**

A process that moves substances without energy expenditure by the organism.

190. **permeability**

Membrane property that allows molecules and ions to pass through.

191. **plasma membrane**

Membrane forming the boundary of a cell.

192. **receptor**

A specific site where a particular substance can bind.

193. **selectively permeable**

A membrane property that allows passage of some substances while preventing passage of others.

194. **solute**

A dissolved substance.

195. **solution**

A liquid containing dissolved substances.

196. **solvent**

A substance in which other substances can dissolve.

197. **surface tension**

Resistance to rupture by the surface film of a liquid.

198. **tonicity**

The degree to which fluid can move into or out of cells.

Carbohydrates / Lipids / Proteins

199. **adenine**

Nitrogenous base classified as a purine found in nucleic acids.

200. **adenosine**

Compound consisting of adenine and ribose.

201. **adenosine diphosphate**

A nucleotide that accepts phosphates in the cell energy cycle.

202. **adenosine triphosphate**

An important energy storage molecule.

203. **aminotransferase**

Enzyme that catalyzes transfer of amino groups from one molecule to another.

204. **beta conformation**

Zig-zag, pleated sheet arrangement of polypeptide chain.

205. **beta oxidation**

Metabolic pathway that oxidizes fatty acids.

206. **beta reduction**

Metabolic pathway that synthesizes fatty acids.

207. **bilayer**

Two rows of amphipathic lipid molecules with hydrocarbon tails facing each other in the middle of the layers.

208. **cholesteryl esterase**

Enzyme that breaks ester bonds between a fatty acid and cholesterol.

209. **citric acid cycle**

Metabolic pathway that oxidizes acetyl-CoA; Krebs cycle; tricarboxylic acid cycle.

210. **conjugated protein**

A protein to which an organic and/or metallic prosthetic group is attached.

211. **cytosine**

Pyrimidine present in nucleotides.

212. **deamination**

Removal of an amino group.

213. **denaturation**

An alteration in the shape and properties of a protein molecule.

214. **denatured protein**

Protein distorted from its native conformation by heat, acid, base or other agent.

215. **deoxyribose**

Five-carbon monosaccharide in DNA.

216. **glycine**

An amino acid with the simplest chemical structure.

217. **histone**

One of five basic (alkaline) proteins associated with eukaryotic chromosomes.

218. **homologous protein**

A protein that has the same function and similar properties in several species.

219. **integral**

Inseparable component of.

220. **isoelectric pH**

The pH at which a solute lacks a net electrical charge.

221. **isoprene**

Recurring hydrocarbon (2-methyl-1,3-butadiene) structural unit of terpenes.

222. **ketogenic amino acid**

Any amino acid whose carbon skeleton can be metabolized to a ketone body.

223. **ketone body**

Acidic molecule that remains from incomplete metabolism of fatty acids.

224. **ketose**

Simple monosaccharide with carbonyl group at a site other than a terminal position.

225. **ketosis**

Accumulation of ketone bodies in blood and urine.

226. **lipoprotein**

A molecule made of lipid and protein.

227. **metabolic water**

Water released from the oxidation of foodstuffs.

228. **monolayer**

Single layer of lipid molecules oriented in a particular way.

229. **mutarotation**

Change in specific rotation of sugar (pyranose/furanose) or glycoside as equilibration of alpha and beta forms occurs.

230. **nonessential amino acid**

Amino acids that an organism can make from other amino acids in proteins.

231.   nucleoside

Molecule containing a purine or pyrimidine and pentose sugar.

232.   nucleotide

A molecule having a nitrogenous base, a 5-carbon sugar, and one or more phosphates.

233.   oligomeric protein

Protein that has two or more polypeptide chains.

234.   oligosaccharide

A group of monosaccharides joined by glycosidic bonds.

235.   pentose

Simple five-carbon sugar.

236.   peptidase

Enzyme that hydrolyzes peptide bonds.

237.   peptide bond

A chemical bond between the amino group of one amino acid and the carboxyl group of another.

238.   pleated sheet

Arrangement of polypeptide chains in a side-by-side sheet.

239.   polypeptide

A chain of amino acids held together by peptide bonds

240.   porphyrin

Nitrogenous compound with four substituted pyrrole rings and usually having a central metal atom.

241.   primary structure of a protein

Amino acid configuration in polypeptide chain.

242. **purine**

A nitrogenous base with two rings found in nucleic acids.

243. **pyrimidine**

A nitrogenous base with one ring found in nucleic acids.

244. **thymine**

Pyrimidine found in DNA that pairs with adenine.

245. **uracil**

A pyrimidine in RNA, which is coded by adenine in DNA.

Enzymes

246. **activation energy**

Energy (kcal) to bring 1 mole of reacting substance to the transition state.

247. **active site**

Surface region of an enzyme that binds substrate and transforms it.

248. **allosteric enzyme**

Regulatory enzyme whose activity can be altered by binding to a specific substance at a noncatalytic site.

249. **allosteric site**

Non-catalytic site to which an activity-altering substance binds.

250. **catalytic site**

Region on an enzyme that participates in catalytic process.

251. **coenzyme**

Simple, nonprotein substance that works with an enzyme.

252. **cofactor**

Small, heat-stable inorganic or organic substance needed for a particular enzyme to act.

253. **competitive inhibition**

Reduction in the rate of an enzyme reaction because substrate and inhibitor compete for active site.

254. **constitutive enzyme**

One of many enzymes always present and active in a cell's main metabolic pathways.

255. **dehydrogenase**

Enzyme that removes hydrogen atoms from substrates.

256. **disulfide bridge**

Covalent link between two polypeptides by cystine residue.

257. **enzyme**

A protein that increases the rate of a chemical reaction in a living organism.

258. **epimerase**

Enzyme that reversibly converts one epimer to another.

259. **homotropic enzyme**

Allosteric enzyme for which the substrate is the modulator.

260. **induced fit**

Modification of enzyme shape to conform to substrate shape.

261. **isomerase**

Enzyme that transforms a substance into its positional isomer.

262. **isozyme**

An isomer of an enzyme; one of two or more forms of an enzyme that catalyze the same reaction.

263. **kinase**

Enzyme that catalyzes phosphorylation of another molecule by ATP.

264. **Lineweaver-Burk equation**

Modification of Michaelis-Menten equation by which Vmax and Km can be more accurately determined.

265. **metalloenzyme**

Enzyme with a prosthetic group consisting of a metal ion.

266. **Michaelis constant (Km)**

Substrate concentration at which enzyme has half its maximum reaction velocity.

267. **Michaelis-Menten equation**

Relationship of reaction velocity to substrate concentration for an enzyme-controlled reaction.

268. **multienzyme system**

Set of enzymes that act sequentially in metabolic pathway.

269. **mutase**

Enzyme that transposes a functional group.

270. **noncompetitive inhibition**

Slowing the rate of a reaction because of binding of an inhibitor to an allosteric site.

271. **optimum pH**

Degree of acidity or alkalinity at which an enzyme has its greatest catalytic activity.

272. **oxygenase**

Enzyme in which oxygen serves as an electron acceptor.

273. **turnover**

Reuse of a substance made available by a catabolic reaction.

General Energy Considerations

274.   activity

Thermodynamic activity or potential of a substance.

275.   activity coefficient

Multiplier by which concentration is multipled to give thermodynamic activity.

276.   aerobe

Organism that requires oxygen.

277.   aerobic

With oxygen.

278.   amphibolic pathway

A metabolic pathway used in both anabolic and catabolic processes.

279.   anabolic

Of anabolism.

280.   anabolism

Synthetic, energy using process.

281.   anaerobe

Organism that does not need and may be harmed by free oxygen.

282.   anaerobic

Lacking oxygen.

283.   anaplerotic reaction

Reaction that can replenish intermediates in citric acid cycle.

284. **ATP synthetase**

Enzyme complex that makes ATP from ADP and phosphate during oxidative phosphorylation.

285. **ATPase**

Enzyme that breaks down ATP to ADP and phosphate usually with transfer of energy to a cellular process.

286. **autotroph**

Organism that synthesizes molecules it needs from simple molecules such as ammonia and carbon dioxide.

287. **auxotroph**

Organism that has lost its ability to make a particular molecule, which it must receive for normal growth.

288. **bioenergetics**

The science of energy changes in living systems.

289. **bond energy**

Energy required to break a chemical bond.

290. **catabolic**

Of catabolism.

291. **catabolism**

Breakdown of molecules that makes energy available.

292. **cellular respiration**

Metabolic processes that yield ATP.

293. **chemiosmotic coupling**

Connection between electron transport and ATP synthesis by way of an electrochemical gradient of hydrogen ions.

294. **chemiosmotic theory**

Explanation of how energy is captured in mitochondria.

295. **common intermediate**

Substance common to two reactions, such as product of one and reactant in next reaction in a pathway.

296. **coupled reaction**

One of a pair of reactions that have a common intermediate and between which energy is transferred.

297. **crossover point**

Point in multienzyme system at which inhibition causes accumulation of intermediates and a decrease in products.

298. **cytochrome**

One of several heme proteins that participate in electron transport during cellular respiration or photosynthesis.

299. **electron acceptor**

Substance that receives electrons in an oxidation-reduction reaction.

300. **electron carrier**

Substance that can transfer electrons from one molecule to another toward free oxygen.

301. **electron donor**

Substance that donates electrons in an oxidation-reduction reaction.

302. **electron transport system**

Enzymes and coenzymes in cristae of mitochondria that move electrons from substrates to oxygen.

303. **endergonic**

Requiring energy, as in a chemical reaction.

304. **energy charge**

Relative degree to which ATP-ADP-AMP system is fully phosphorylated.

305. **energy coupling**

Energy transfer from one process to another.

306. enthalpy

Heat in a system.

307. entropy

Tendency toward chaos or disorder.

308. equilibrium

State of a system having minimal free energy and not undergoing change.

309. equilibrium constant

A constant value that relates concentrations of reactants and products of a reaction under given conditions.

310. exergonic

Releasing energy, as in a chemical reaction.

311. facultative

Able to live with or without free oxygen.

312. fermentation

Anaerobic breakdown of glucose or other nutrient.

313. flavin adenine dinucleotide (FAD)

A coenzyme that carries hydrogen.

314. gluconeogenesis

Metabolic pathway that makes glucose from noncarbohydrate substances.

315. glucose sparing

Metabolism of fats by many cells that conserves glucose in blood for transport to cells that cannot metabolize fats.

316. glycogenesis

Metabolic pathway for glycogen synthesis.

317. **glycogenolysis**

Metabolic pathway for glycogen breakdown.

318. **glycolysis**

Metabolic pathway for breakdown of glucose to pyruvic acid.

319. **guanidine triphosphate (GTP)**

An energy storage molecule.

320. **intermediary metabolism**

Enzyme-controlled reactions that take chemical energy from nutrients and use it to make cell components.

321. **intermediate**

A molecule produced within a metabolic pathway.

322. **irreversible process**

A process in which entropy increases.

323. **isothermal process**

Process that occurs at constant temperature.

324. **kinetic**

Pertaining to the energy of motion.

325. **low-energy phosphate**

Phosphorylated substance that has relatively little free energy.

326. **metabolism**

All chemical reactions in a living organism.

327. **metabolite**

Intermediate substance in metabolic reactions.

328. **nicotinamide adenine dinucleotide (NAD)**

A coenzyme that transports hydrogen atoms or electrons in oxidation-reduction reactions.

329. **open system**

System that exchanges matter and energy with surroundings.

330. **orthophosphate cleavage**

Enzymatic cleavage of ATP yielding ADP and phosphate, and usually coupled to a process that requires energy.

331. **oxidative phosphorylation**

Capture of energy in ATP during oxidative metabolism.

332. **pentose phosphate pathway**

Metabolic pathway that produces five-carbon sugars and reduced NADP.

333. **phosphogluconate pathway**

Metabolic pathway that uses glucose-6-p to reduce NADP and make pentoses and other products.

334. **phosphorylation**

Binding of a phosphate group to a molecule.

335. **potential energy**

Energy due to position and capable of being released, as in a rock at the top of a hill.

336. **specific heat**

The amount of heat needed to increase the temperature of a specific volume of substance one degree Celsius.

337. **uridine triphosphate (UTP)**

A high energy molecule.

Photosynthesis

338.   **absorption spectrum**

Relative degree to which a pigment absorbs different wavelengths of light.

339.   **accessory pigment**

Molecule with light-gathering properties that augments chlorophyll and may give color to plant tissue.

340.   **action spectrum**

Relative degree to which different wavelengths of light affect a light-dependent process.

341.   **bacteriochlorophyll**

A kind of chlorophyll found in bacteria that can receive electrons from sources other than water.

342.   **bundle sheath cell**

Internal leaf cell near vascular bundle where carbon metabolism by way of the Calvin cycle occurs.

343.   **C-four plant**

Plant in which carbon is fixed in 4-carbon compound.

344.   **C-three plant**

Plant in which carbon is fixed in 3-carbon compound.

345.   **Calvin cycle**

Series of reactions that fix carbon from carbon dioxide and subsequently synthesize sugars.

346.   **carotenoid**

Yellow-orange plant pigment that absorbs violet to green light.

347.   **CF-one**

Part of ATP synthetase in chloroplast that generates ATP.

348. **CF-zero**

Part of ATP synthetase in thylakoid membrane that moves protons.

349. **chlorophyll**

A green pigment capable of capturing light energy.

350. **chloroplast**

Photosynthetic organelle that contains chlorphyll and is found in eukaryotic cells.

351. **cyclic electron flow**

Transfer of electrons from photosystem I with energy made available for ATP synthesis.

352. **cyclic photophosphorylation**

The capture of energy in chloroplasts with the formation of ATP through the activity of the cell's cytochrome system.

353. **dark reaction**

A part of photosynthesis that can occur in either light or dark and that transfers energy from light reactions.

354. **Emerson enhancement effect**

Greater photosynthetic activity produced with red light of two slightly different wavelengths.

355. **excited state**

Energy-rich state of an atom or molecule as a result of absorbing light energy.

356. **fluorescence**

Light emission by excited molecules as they return to their ground state.

357. **glycolate pathway**

Photosynthetic reactions by which phosphoglycolate is made in chloroplasts.

358. **granum**

A stack of thylakoids within a chloroplast.

359. **ground state**

Stable, non-excited state of an atom or molecule that can be excited by light energy.

360. **Hill reaction**

Photoredution of electron acceptor and oxygen release by a chloroplast in the absence of carbon dioxide.

361. **intrathylakoid space**

Space within membranes of thylakoids and lamellae of stroma.

362. **light reaction**

Events in photosynthesis that capture energy and occur in light.

363. **mesophyll cell**

Outer cell in C-four plant leaf; site of carbon fixation.

364. **nicotinamide adenine dinucleotide phosphate**

Coenzyme that transfers electrons in the Calvin cycle and other metabolic pathways.

365. **noncyclic electron flow**

Electron flow from water to NADP using light for energy.

366. **noncyclic photophosphorylation**

The capture of energy in chloroplasts with the formation of ATP and reduced NADP.

367. **oxygenic photoautotroph**

Organism in which water is the electron donor in photosynthesis.

368. **P/O ratio**

Ratio of ATP molecules generated per oxygen atom reduced.

369. **photochemical reduction**

Transfer of light-excited electrons from one molecule to another.

370. photoexcitation

Excitation of an electron by absorption of light energy.

371. photolysis

Use of light energy to oxidatively split water molecules.

372. photon

The smallest unit of light energy.

373. photophosphorylation

The addition of phosphate groups and high energy bonds to a molecule during the capture of light energy.

374. photoreduction

Use of light energy to generate NADPH by transfer of excited electrons from chlorophyll to NADP via electron carriers.

375. photorespiration

Oxygen use in light by temperate zone plants through oxidation of phosphoglycolate.

376. photosynthesis

The process by which organisms capture light energy from the environment and store it in a usable form.

377. photosynthetic phosphorylation

Enzymatic production of ATP from ADP with electron transport using light energy in photosynthetic organisms.

378. photosynthetic unit

Group of up to 300 chlorophyll molecules of which only a few molecules participate in photochemical reactions.

379. photosystem

Functional system of chlorophyll and other pigments embedded in thylakoid membrane.

380. photosystem I

System in which chlorophyll absorbs 700nm red light maximally after which electrons reduce NADP to NADPH.

381. **photosystem II**

System in which chlorophyll absorbs 680nm red light maximally after which electrons donated by water are excited.

382. **phycobilin**

Pigment in red and blue-green algae that absorbs green-to-orange light and gives algae their color.

383. **pigment**

Substance that absorbs some light and reflects light of wavelengths corresponding to the color perceived.

384. **plastocyanin**

Protein containing copper that donates electrons to chlorophyll P700 of photosystem I in light reactions.

385. **plastoquinone**

Molecule with quinone component that participates in transfer of electrons between photosystems I and II.

386. **quantum**

Fundamental unit of light energy.

387. **quantum requirement**

Amount of light energy in photons needed for specific change.

388. **reaction center**

Part of photosynthetic unit having chlorophyll molecules that intiate electron transfer.

389. **stoma**

A pore in lower leaf epidermis through which gases diffuse into and out of mesophyll spaces.

390. **thylakoid**

Parallel flattened sacs that form part of the membrane structure of a chloroplast.

Nucleic Acid Metabolism / DNA Replication

391. **antiparallel**

Opposite in orientation or polarity.

392. **B-DNA**

Right-handed DNA helix.

393. **bacteriophage**

Virus that can infect a bacterium.

394. **base pair**

Nucleotides in different chains of nucleic acids held together by hydrogen bonds.

395. **capsid**

Protein coat of a virus.

396. **central dogma**

Principle that information flows from DNA to RNA to protein.

397. **Chargaff's rules**

In double stranded DNA, the number of adenines and thymines and the number of cytosines and guanines are equal.

398. **chimeric DNA**

Recombinant DNA that has genes from two different species.

399. **complementary base pairing**

Bonding between certain bases in nucleic acid strands.

400. **constitutive heterochromatin**

Permanently condensed genetically inactive chromosomal region.

401. **deoxyribonuclease**

An enzyme that digests DNA.

402. **deoxyribonucleic acid (DNA)**

A nucleic acid in chromosomes that directs protein synthesis and transmits genetic information to a new generation.

403. **DNA chimera**

DNA containing genes from two different species.

404. **DNA glycosidase**

Enzyme that finds and removes deaminated bases from DNA.

405. **DNA ligase**

Enzyme that forms phosphodiester bond between antiparallel ends of DNA segment while it is base-paired to a template.

406. **DNA replicase system**

Enzyme complex and proteins needed in DNA replication.

407. **DNA replication**

Synthesis of new DNA according to information in an existing DNA template.

408. **double helix**

Natural conformation of complementary antiparallel chains of DNA.

409. **endonuclease**

Enzyme that can hydrolyze phosphodiester bonds at points in molecule other than ends.

410. **euchromatin**

Diffuse, uncondensed, active chromatin.

411. **exonuclease**

Enzyme that hydrolyzes phosphodiester bonds at ends of nucleic acid chains.

412. **genome**

An organism's whole complement of DNA.

413. **helix**

Spiral shape of DNA and some other biopolymers.

414. **insertion**

Mutation in which an extra base or mutagen has been placed between two successive bases in DNA.

415. **insertion sequence**

Particular base sequences at either end of a transposable DNA segment.

416. **insertional mutagenesis**

Changing of transcriptional activity of nearby host genes by insertion of viral DNA.

417. **major groove**

Larger of grooves in double-helix DNA.

418. **minor groove**

Smaller of grooves between DNA strands in double helix.

419. **mutation**

A change in genetic information.

420. **negative supercoil**

Coil in circular DNA by left-handed twist of relaxed molecule.

421. **nonhistone chromosomal protein**

Any of several acidic proteins found in small amounts in eukaryotic chromosomes.

422. **nonreiterated sequence**

DNA sequence found only once per haploid genome.

423. nonsense mutation

Mutation that prematurely terminates a polypeptide chain.

424. nuclease

Enzyme that can hydrolyze links between nucleotides of a nucleic acid.

425. nucleoside diphosphate sugar

Coenzyme-like sugar molecule carrier in synthesis of polysaccharides and sugar derivatives.

426. nucleoside diphosphokinase

Enzyme that moves a terminal phosphate from a triphosphate to a monophosphate.

427. nucleosome

Structural unit of a chromosome containing 200 DNA base pairs associated with an octamer of histone proteins.

428. packing ratio

Ratio of the length of a DNA molecule to the length of a chromosome into which it is packed.

429. palindromic

Reading the same forward and backward.

430. plasmid

Extrachromosomal DNA that can replicate independently of a bacterial chromosome.

431. polarity

Difference between 3' and 5' ends of nucleotide chains in a nucleic acid.

432. positive supercoil

Coil in circular DNA formed by right-handed twist of a relaxed molecule.

433. reiterated sequence

DNA sequence found in multiple copies in a haploid genome.

434. **relaxed state**

Circular DNA that lacks supercoils.

435. **repair endonuclease**

Enzyme that detects missing bases in DNA and breaks bonds at such sites so other repair enzymes can act.

436. **repair synthesis**

Removal of defective DNA segments and replacement with normal ones.

437. **replication**

Duplication.

438. **replicon**

Self-replication DNA unit that includes site at which replication began.

439. **restriction enzyme**

An endonuclease that cuts double stranded DNA at sites having specific nucleotide sequences.

440. **restriction site**

The site at which an endonuclease acts.

441. **spacer sequence**

Sequence of RNA nucleotides excised during processing.

442. **supercoil**

Twist in circular DNA that makes helix coil on itself.

443. **topoisomerase**

Enzyme that converts relaxed DNA to supercoiled DNA.

Protein Synthesis

444. actinomycin D

An inhibitor of RNA synthesis.

445. amino acid activation

Esterification of carboxyl group with appropriate tRNA using energy from ATP.

446. aminoacyl-tRNA

Aminoacyl ester of tRNA.

447. aminoacyl-tRNA synthetase

Enzyme that catalyzes synthesis of aminoacyl-tRNA.

448. anticodon

A three-base sequence of transfer RNA that fits with a particular codon on messenger RNA.

449. base pairing

Bonding of complementary purines and pyrimidines in double-stranded nucleic acids.

450. catabolite activator protein

Substance that can initiate transcription of genes to make enzymes to metabolize other nutrient when glucose is used.

451. chimeric protein

Protein synthesized from information in chimeric DNA.

452. codon

A three-base sequence in messenger RNA derived from DNA and specifying amino acid placement in a protein.

453. cohesive end

Single-stranded DNA fragment from cleavage by a restriction enzyme that can attach to another similar fragment.

454. colinearity

Direct correspondence between nucleotide sequences in DNA and mRNA and the amino acids in a protein.

455. complementary DNA

DNA that is complementary to mRNA made by reverse transcriptase.

456. consensus sequence

Nucleotide order in highly conserved segment of DNA.

457. coordinate induction

Turning on of a set of enzymes by a single inducer.

458. core particle

Histone octamer having DNA wound around it.

459. degenerate code

Code having element in one language expressed by more than one element in another language.

460. depurination

Removal of a purine.

461. DNA polymerase

An enzyme that increases chain length in DNA synthesis.

462. elongation factor

Protein required for elongation of polypeptide chain.

463. end-product inhibition

Slowing of an allosteric enzyme at the beginning of a pathway by its end product.

464. enzyme repression

Inhibition of synthesis by accumulation of molecules being synthesized.

465. **exon**

DNA segment that is transcribed into mRNA and codes for a particular domain of a protein in eukaryotic cell.

466. **facultative heterochromatin**

Region of chromosome that has been specifically inactivated in a particular kind of cell.

467. **genetic code**

The three-base sequences in messenger RNA derived from a DNA template that determine amino acid order in proteins.

468. **gout**

A joint inflammation due to uric acid accumulation.

469. **guanine**

A purine found in nucleotides of nucleic acids.

470. **induced enzyme**

Enzyme that a cell makes only when its substrate is present.

471. **inducer**

Molecule (usually substrate) that can induce synthesis of an enzyme.

472. **initiation codon**

Codon that codes for first amino acid in a protein.

473. **initiation complex**

Ribosome-mRNA complex and initiating tRNA needed to elongate a polypeptide.

474. **initiation factor**

One of several specific proteins needed to start polypeptide synthesis on a ribosome.

475. **intron**

DNA that is transcribed but excised from mRNA before it is used to direct polypeptide synthesis.

476. **messenger RNA (mRNA)**

A nucleic acid that carries information as codons for protein synthesis.

477. **modulator**

Metabolite that acts at an allosteric site to alter properties of a regulatory enzyme.

478. **nitrogen balance**

The situation in which the quantity of nitrogen entering the equals the quantity of nitrogen leaving the body.

479. **nonheme-iron protein**

Protein that contains iron but no porphyrin.

480. **nonsense codon**

Codon that codes for termination of chain and not for any amino acid.

481. **operator**

DNA region that interacts with a repressor protein and determines expression of a gene (or group of genes).

482. **operon**

Set of related genes and their operator and promotor sequences.

483. **polynucleotide**

Sequence of nucleotides joined by pentoses by phosphodiester bonds.

484. **polyribosome**

A mRNA complex with two or more ribosomes.

485. **posttranslational modification**

Processing of a polypeptide after synthesis.

486. **puromycin**

Antibiotic that inhibits polypeptide synthesis by binding to chain and preventing elongation.

487. **releasing factor**

Cytosolic protein needed to release a newly synthesized protein from its ribosome.

488. **ribonucleic acid (RNA)**

A nucleic acid made from information in DNA that is involved in protein synthesis.

489. **sequence homology**

Same order of amino acids in different proteins.

490. **structural gene**

Gene that codes for the primary structure of a protein.

491. **termination codon**

Three-base set in mRNA that stops elongation of a polypeptide; UAA, UAG, or UGA.

492. **termination sequence**

DNA sequence located at the end of a transcriptional unit that stops transcription.

493. **transamination**

Transfer of an amino group from one molecule to another.

494. **transcription**

The transfer of coded genetic information from DNA to mRNA.

495. **transcriptional control**

Control of protein synthesis by controlling mRNA synthesis.

496. **transfer RNA (tRNA)**

RNA that carries amino acids to specific sites in a growing peptide chain.

497. **translation**

The process by which mRNA codons are used to determine the sequence of amino acids in a protein.

498. **translational control**

Control of protein synthesis by controlling the rate of reactions on a ribosome.

499. **wobble**

Base pairing in which several codon/anticodon combinations place the same amino acid in a protein.

500. **xanthine**

A purine that inhibits cAMP breakdown.

Digestion / Transport of Nutrients

501. **absorption**

Movement of nutrients from digestive tract into blood.

502. **absorptive**

Concerning absorption.

503. **aminopeptidase**

An enzyme that digests peptides from the amino end.

504. **amylase**

An enzyme that digests starch.

505. **argentaffin cells**

Stomach lining cells that secrete histamine and serotonin.

506. **bile**

Liver secretion that aids in digestion by emulsifying fats.

507. **bile salt**

Steroid with amphipathic and detergent properties that emulsifies lipids.

508. **brown fat**

Fat with a high energy content deposited around organs in newborn infants.

509. **carboxypeptidase**

A proteolytic enzyme that digests peptides from the carboxyl end.

510. **CCK (cholecystokinin-pancreozymin)**

An enteric hormone that stimulates the gallbladder to release bile and the pancreas to secrete enzymes.

511. **chenodeoxycholic acid**

A bile acid.

512. **chief cell**

A gastric gland cell that secretes pepsin into the stomach.

513. **cholic acid**

A bile acid.

514. **chylomicron**

A particle made of lipids and protein in the intestinal mucosa and released into lacteals.

515. **chyme**

Semiliquid, partially digested food leaving the stomach.

516. **chymotrypsin**

A proteolytic enzyme from the pancreas that hydrolyzes proteins into polypeptides and amino acids.

517. **chymotrypsinogen**

Inactive chymotrypsin.

518. **co-transport**

Movement of two substances across a membrane by the same carrier.

519. **colipase**

An enzyme that assists a lipase.

520. **digestion**

Breakdown of large molecules into smaller ones.

521. **dipeptidase**

An enzyme that digests dipeptides to amino acids.

522. **dipeptide**

A molecule of two amino acids held together by a peptide bond.

523. **enterokinase**

A proteolytic enzyme from the intestinal mucosa.

524. **gliadin**

A protein in wheat and some other grains that damages the intestinal mucosa in sensitive individuals.

525. **gluten**

A protein containing gliadin found in wheat and some other grains.

526. **hepatitis**

Liver inflammation.

527. **hydrochloric acid**

Acid released from gastric mucosa needed for protein digestion.

528. **intrinsic factor**

Substance secreted by the gastric mucosa required for the transport and absorption of vitamin B12.

529. **lactase**

An enzyme that digests lactose.

530. **lacteal**

Lymph vessel in a villus of the small intestine.

531. **lipase**

An enzyme that breaks down lipids.

532. **maltase**

Enzyme that digests maltose, a disaccharide from starch.

533. **micelle**

A small fat droplet in chyme.

534. **microvillus**

A cytoplasmic projection of surface membrane of intestinal epithelial cells.

535. **pancreas**

A digestive gland that secretes enzymes and hormones.

536. **pancreatic**

Of the pancreas.

537. **pancreatic juice**

Fluid from pancreas containing bicarbonate and digestive enzymes for all types of food molecules.

538. **pancreatitis**

Inflammation of the pancreas.

539. **pepsin**

An enzyme that starts breakdown of protein in the stomach.

540. **procarboxypeptidase**

Inactive carboxypeptidase.

541. **ribonuclease**

Enzyme that digests RNA.

542. **secretagogue**

Substance that stimulates secretion of digestive juices.

543. **secretin**

A hormone from the intestinal mucosa that stimulates secretion of bile and pancreatic fluid.

544.   **sprue**

Inflammation and partial destruction of the gastrointestinal mucosa.

545.   **sucrase**

An enzyme that digests sucrose.

546.   **taurine**

An amino acid derivative that conjugates with bile acids.

547.   **trypsin**

A proteolytic enzyme from the pancreas that digests proteins into peptone.

548.   **trypsinogen**

Inactive trypsin.

549.   **ulcer**

Erosion of a mucous membrane, such as gastric or duodenal mucosa.

550.   **ulcerative**

Of an ulcer.

551.   **villikinin**

A mucosal hormone that causes villi to move.

552.   **villus**

Vascular tuft.

Nutrition

553. **alcoholism**

Disease of repetitive, excessive alcohol intake, which leads to severe problems in daily living.

554. **anorexia**

Lack of appetite.

555. **anorexia nervosa**

A serious neurological disorder in which a person loses weight and becomes emaciated.

556. **ascorbic acid**

Vitamin C.

557. **basal metabolic rate (BMR)**

Rate of energy use that maintains body processes in an awake, resting individual.

558. **basal metabolism**

The process of using energy from nutrients to maintain life in an awake, resting state.

559. **biotin**

Vitamin required for fat synthesis.

560. **bulimia**

Binge eating, usually followed by self induced vomiting.

561. **calciferol**

Steroid having vitamin D activity.

562. **calorie**

Quantity of heat to raise the temperature of one gram of water one degree Celsius.

563. **carbohydrate-craving obesity**

Excessive body weight caused by a persistent desire for sugars and starches.

564. **carotene**

Yellow substance that usually has vitamin A activity.

565. **coenzyme A**

Molecule containing pantothenic acid that carries acyl groups in certain reactions.

566. **core body temperature**

Temperature deep within the body.

567. **cyanocobalamin**

Vitamin transported by intrinsic factor and required for normal cell division.

568. **diabetes mellitus**

A disorder due to lack or inactivity of insulin that allows glucose to accumulate in the blood and urine.

569. **essential amino acid**

Amino acid required in the diet because the body cannot make it.

570. **essential fatty acid**

Fatty acid required in the diet because the body cannot make it.

571. **estrogen**

One of several active molecules that stimulate development of female organs and secondary sexual characteristics.

572. **folacin**

Vitamin needed to help transfer single carbon groups.

573. **growth hormone-hypothalamic mechanism**

A metabolic regulatory mechanism involving pituitary-related hormones.

574. **hyperglycemia**

Abnormally high blood glucose concentration.

575. **hyperthermia**

An abnormally high body temperature.

576. **hypoglycemia**

Abnormally low blood glucose concentration.

577. **hypoglycemic**

An agent that lowers the blood glucose concentration.

578. **hypothermia**

An abnormally low body temperature.

579. **insulin shock**

Unconsciousness due to an insulin overdose that suddenly lowers blood glucose.

580. **insulin-glucagon mechanism**

A metabolic regulatory mechanism involving pancreatic hormones.

581. **kilocalorie**

Heat required to raise the temperature of one kilogram of water one degree Celsius.

582. **kwashiorkor**

Malnutrition involving protein deficiency, usually in young children.

583. **lipoic acid**

Vitamin for some microorganisms that serves as a hydrogen carrier.

584. **malnutrition**

Ill health caused by an inadequate diet.

585. **marasmus**

Malnutrition to the degree of near-starvation.

586. **metabolic rate**

Rate at which nutrients are oxidized.

587. **mineral**

Inorganic substance.

588. **net protein utilization**

Proportion of protein eaten that is actually used by cells.

589. **niacin**

B vitamin used to synthesize the coenzyme NAD.

590. **nutrition**

The act of providing substances needed for good health through food ingestion.

591. **obesity**

Condition of having excessive body fat.

592. **pantothenic acid**

B vitamin used to synthesize coenzyme A.

593. **pica**

Craving for nonfood substances.

594. **polydipsia**

Excessive fluid intake.

595. **polyphagia**

Excessive eating.

596. **polyuria**

Excessive urine production.

597. **postabsorptive**

Related to metabolism after food from a meal is completely absorbed.

598. **respiratory quotient (RQ)**

Ratio of carbon dioxide released to oxygen consumed.

599. **riboflavin**

Heat-labile B vitamin used to synthesize the coenzyme FAD.

600. **scurvy**

A disease due to a vitamin C deficiency.

601. **selenosis**

A disorder caused by toxic levels of selenium in the body.

602. **thiamine**

A water-soluble B vitamin used to synthesize cocarboxylase.

603. **tocopheral**

A substance having vitamin E activity.

604. **total parental nutrition (TPN)**

Process of giving all required nutrients by a route other than the digestive tract.

605. **vitamin A**

Vitamin needed to synthesize visual pigments and maintain epithelial cells.

606. **vitamin D**

A vitamin that facilitates calcium absorption.

607. **vitamin E**

Vitamin that acts as an antioxidant.

608. **vitamin K**

Vitamin needed for synthesis of some blood clotting factors.

Hormones

609.   **adrenal**

Above the kidney; a gland lying superior to the kidney.

610.   **adrenalin**

A hormone secreted by the adrenal medulla; epinephrine.

611.   **adrenocorticotropic hormone**

A hormone that stimulates the adrenal cortex to secrete hormones.

612.   **alarm reaction**

The body's characteristic response to a stressful situation.

613.   **aldosterone**

An adrenocortical hormone that increases reabsorption of sodium.

614.   **anabolic steroid**

A synthetic hormone that increases muscle size.

615.   **androgen**

A molecule with male hormone activity.

616.   **angiotensin**

A substance that causes vasoconstriction and aldosterone release.

617.   **angiotensinogen**

Inactive precursor of angiotensin.

618.   **antidiuretic hormone (ADH)**

A hypothalamic hormone stored in the posterior pituitary gland that stimulates water conservation by the kidneys.

619. **atrial natriuretic hormone (ANH)**

A substance secreted by the atria that accelerates sodium excretion in the kidneys.

620. **beta lipotropin**

A molecule from which endorphins are derived.

621. **calcitonin**

A hormone that lowers blood calcium.

622. **calmodulin**

An intracellular calcium carrier molecule.

623. **corticosterone**

A steroid hormone from the adrenal cortex.

624. **cortisol**

An adrenocortical hormone that helps regulate carbohydrate metabolism and counteracts inflammation.

625. **cyclic AMP**

Intracellular second messenger that mediates effects of hormones at receptors on the cell membrane.

626. **diabetes insipidus**

A disorder in which a lack of antidiuretic hormone leads to production of large quantities of dilute urine.

627. **diffuse endocrine system (DES)**

Assortment of hormone-secreting cells in various locations throughout the body.

628. **dynorphin**

A polypeptide related to an enkephalin.

629. **endocrine**

Concerning a ductless gland.

630. **endorphin**

A peptide that binds to opiate receptors in the brain.

631. **enkephalin**

A peptide derived from endorphin that binds to opiate receptors in the brain.

632. **enteric**

Of the intestine.

633. **epinephrine**

Main hormone from the adrenal medulla.

634. **estradiol**

Primary human estrogen.

635. **eustress**

A productive kind of stress.

636. **exhaustion stage**

Stage of stress at which the body has failed to cope.

637. **follicle-stimulating hormone (FSH)**

A hormone that stimulates maturation of ova and sperm.

638. **gastrin**

A hormone from the stomach lining that circulates in the blood and stimulates HCl secretion.

639. **general adaptation syndrome**

A set of changes that occur in animals under stress.

640. **glucagon**

A hormone that raises blood glucose.

641. **glucocorticoid**

A hormone that helps to regulate carbohydrate metabolism.

642. **gonadotropin**

Hormone from the anterior pituitary gland that stimulates gonads.

643. **growth hormone**

Hormone from the anterior pituitary gland that stimulates growth and maintains adult body size.

644. **hormone**

A regulatory substance from an endocrine cell that is transported in the blood to its target cells.

645. **human chorionic gonadotropin**

A placental hormone that stimulates the corpus luteum to secrete hormones.

646. **hypophysis**

The pituitary gland.

647. **insulin**

A hormone from the pancreas that causes cells to take in glucose and stimulates protein synthesis.

648. **islet of Langerhans**

Cluster of hormone-secreting cells in the pancreas.

649. **lutein**

A yellow pigment.

650. **luteinizing hormone (LH)**

A hormone that helps to cause ovulation and other reproductive processes.

651. **mineralocorticoid**

A hormone that regulates mineral metabolism.

652.   natriuresis

Stimulation of sodium excretion.

653.   neurohypophysis

The posterior, neural part of the pituitary gland.

654.   oxytocin

A hormone from the hypothalamus that stimulates uterine contractions and milk let down.

655.   parathormone

A hormone from the parathyroid gland that decreases blood calcium.

656.   parathyroid glands

Glands imbedded in the thyroid gland that produce a hormone important in calcium metabolism.

657.   pituitary gland

Neuroendocrine gland at base of brain, which provides connection between nervous and endocrine functions.

658.   progesterone

A hormone that helps to maintain pregnancy.

659.   prolactin

A hormone that stimulates milk secretion.

660.   prostaglandin

A substance derived from the fatty acid arachidonic acid that acts over short distances as a chemical messenger.

661.   puberty

A period during which sexual maturity is achieved.

662.   relaxin

A hormone from the corpus luteum released during pregnancy.

663. **resistance stage**

A period during which the body successfully copes with stress.

664. **second messenger**

Intracellular substance that relays a signal from an extracellular substance bound to a membrane receptor.

665. **somatostatin**

Growth-hormone inhibiting hormone.

666. **somatotropin**

Growth hormone.

667. **stress**

A condition produced by a variety of injurious agents that affects many body systems.

668. **stressor**

An agent or event that produces stress.

669. **target cell**

A cell that can respond to a certain hormone.

670. **testosterone**

A male hormone.

671. **thymosin**

A hormone secreted by the thymus gland.

672. **thymus gland**

A gland that processes and activates T lymphocytes before it regresses during puberty.

673. **thyroid gland**

A gland in the throat that produces hormones important in regulating the metabolic rate.

674.   **thyroid-stimulating hormone (TSH)**

A hormone that stimulates hormone secretion by the thyroid gland.

675.   **tropic**

Influencing another organ or process.

676.   **vasopressin**

Antidiuretic hormone.

Gas Exchange and Respiration

677. anoxia

Lack of oxygen.

678. aortic body

Receptor in aortic arch that detects changes in blood gases and blood pH.

679. asthma

A disorder in which constriction of bronchioles causes difficulty in breathing.

680. atelectasis

Collapse of lung alveoli or their failure to expand at birth.

681. Bohr effect

Tendency of a high oxygen concentration in the lungs to facilitate release of carbon dioxide from hemoglobin.

682. Boyle's law

Pressure exerted by a gas is inversely proportional to its volume.

683. carbaminohemoglobin

Hemoglobin to which carbon dioxide is bound.

684. Dalton's law

Each gas in a mixture exerts a partial pressure that is independent of other gases.

685. emphysema

A disorder characterized by destruction or dilation of walls of alveoli.

686. heme

An iron-containing pigment in hemoglobin that binds oxygen.

687. hemoglobin

The oxygen-carrying protein in erythrocytes.

688. hyaline membrane disease (HMD)

A lack of surfactant in newborns that allows alveoli to collapse.

689. hypertension

Excessively high blood pressure.

690. hypoxia

Oxygen deficiency in cells.

691. oxyhemoglobin

Hemoglobin to which oxygen is bound.

692. partial pressure

Pressure exerted by one gas in a mixture of gases.

693. perfluorocarbon

An oxygen-binding substance used in synthetic blood.

694. respiration

The processes of ventilation (breathing) and gas exchange.

695. respiratory distress syndrome

Labored breathing and impaired gas exchange because of surfactant deficiency.

696. surfactant

A phospholipid that reduces surface tension.

Body Fluids and Circulation

697. **albumin**

A small protein made in the liver and released into blood.

698. **anemia**

A hemoglobin deficiency associated with too few erythrocytes or poorly functioning ones.

699. **anticoagulant**

A substance the prevents blood from clotting.

700. **antithrombin**

Substance in plasma that inhibits coagulation by neutralizing thrombin.

701. **apoferritin**

A protein that carries iron.

702. **atherosclerosis**

Obstruction of arteries by hardened plaque deposits.

703. **bilirubin**

A red bile pigment from hemoglobin breakdown.

704. **biliverdin**

A green bile pigment from hemoglobin breakdown.

705. **blood**

Fluid pumped by the heart through a closed system of vessels.

706. **blood-brain barrier**

A specialized capillary structure that limits movement of substances from blood into brain tissue.

707. **bradykinin**

A polypeptide with a potent vasodilating action.

708. **cardiopulmonary resuscitation (CPR)**

A method for maintaining blood flow and gas exchange in a person with no heart beat and no breathing.

709. **cardiovascular**

Of the heart and blood vessels.

710. **cerebrospinal fluid**

A clear fluid in spaces within and around the central nervous system.

711. **ceruloplasmin**

A blood protein that transports copper.

712. **clot retraction**

Shrinkage of a blood clot.

713. **coumarin**

An anticoagulant that blocks synthesis of some clotting factors in the liver.

714. **cross-matching**

Comparison of donor and prospective recipient bloods to detect possibilities of agglutination.

715. **endolymph**

Fluid located in the membranous labyrinth of the inner ear.

716. **erythrocyte**

Red blood cell.

717. **erythropoiesis**

Process of red blood cell formation.

718. erythropoietin

A substance from the kidney that stimulates erythropoiesis.

719. ferritin

A molecule made up of the protein apoferritin and iron.

720. fibrin

A fibrous protein that forms a network in a blood clot.

721. fibrinogen

An inactive precursor of the protein fibrin.

722. gastroferrin

An iron-binding protein that transports iron from the stomach lumen to mucosal cells.

723. globin

A globular protein found in hemoglobin and certain other biological molecules.

724. globulin

A globular protein, including many in the plasma.

725. glomerulus

A capillary tuft surrounded by a glomerular capsule.

726. hematocrit

The proportion of erythrocytes in a volume of blood.

727. hematopoiesis

The formation of blood cells; hemopoiesis.

728. hemolysis

Breakdown of erythrocytes with hemoglobin release.

729. **hemolytic**

Concerning hemolysis.

730. **hemopoietic**

Related to hemopoiesis.

731. **hemorrhage**

Loss of a significant volume of blood.

732. **hemosiderin**

A molecule that binds and stores iron.

733. **hemostasis**

The arrest of bleeding.

734. **heparin**

An anticoagulant synthesized in several body tissues.

735. **hirudin**

An anticoagulant secreted by leeches.

736. **hydrocephalus**

Excessive cerebrospinal fluid in brain ventricles; water on the brain.

737. **ischemia**

Reduction in blood flow to an area.

738. **jaundice**

Yellowish tone to skin and membranes caused by excess bile pigments in the blood.

739. **lymph**

Interstitial fluid in a lymphatic vessel.

740. perfusion

The flow of blood through vessels.

741. pernicious anemia

Anemia due to a lack of intrinsic factor and therefore vitamin B12.

742. plasma

The fluid part of blood including inactive clotting factors.

743. plasma protein

Protein found in blood plasma.

744. plasmin

An enzyme built into blood clots as they form that gradually dissolves them.

745. plasminogen

Inactive plasmin.

746. platelet

A megakaryocyte fragment in blood that participates in blood clotting reactions.

747. prothrombin

Inactive thrombin.

748. serum

The fluid part of blood after removal of formed elements and clotting factors.

749. sickle cell anemia

An inherited anemia in which erythrocytes sickle under low oxygen conditions.

750. spectrin

A protein that maintains the flexibility of erythrocyte membranes.

751. streptokinase

An enzyme that digests blood clots used to treat coronary occlusion.

752. thalassemia

An anemia caused by a deficiency of alpha or beta chains hemoglobin.

753. thrombin

An enzyme that activates fibrinogen to fibrin in the blood clotting mechanism.

754. tissue plasminogen activator (tPA)

A substance secreted by many tissues that activates plasminogen to plasmin.

755. tissue thromboplastin

A substance from injured tissue that initiates extrinsic blood clotting.

756. transferrin

An iron-transport protein in plasma.

Skin and Connective Tissues

757. **acid mucopolysaccharide**

Acidic sugar polymer in mucous secretions of higher animals.

758. **adipose**

Pertaining to fat.

759. **cartilage**

A firm, resilient, flexible connective tissue.

760. **elastin**

Main protein in elastic fibers of connective tissues.

761. **eleidin**

A keratin precursor found in the stratum lucidum.

762. **fibroblast**

A connective tissue cell that makes fibers and ground substance.

763. **goblet cell**

Single-celled gland that produces mucus.

764. **ground substance**

A glycoprotein deposited among fibers of connective tissue.

765. **hydroxyapatite**

A mineral that makes up the bulk of bone.

766. **keratin**

Water-insoluble protein in skin, hair, and nails.

767.   keratohyalin

A translucent substance in skin.

768.   melanin

A dark brown pigment of hair and skin.

769.   mucin

A glycoprotein in ground substance and mucous secretions.

770.   mucopolysaccharide

Sugar polymer found in mucous secretions.

771.   mucoprotein

Protein conjugated to a mucopolysaccharide; proteoglycan.

772.   mucous

Pertaining to mucus.

773.   mucus

Thick secretion from a goblet cell.

774.   osteoporosis

Abnormal porousness of bone, which makes it fracture-prone.

775.   rickets

A failure of bones to harden in childhood because of a calcium deficiency.

776.   sebaceous

Of sebum.

777.   sebum

A substance containing oils and epithelial cell debris from sebaceous glands.

778. **sudoriferous**

Secreting sweat.

779. **sudoriferous gland**

Sweat gland.

Nervous System

780. acetylcholine

A neurotransmitter released by many axons, especially those that control skeletal muscles.

781. adrenergic

Concerning a neuron that releases norepinephrine (adrenalin).

782. Alzheimer's disease

A degenerative neurological disorder associated with memory loss and behavioral changes.

783. anesthetic

Agent that produces temporary loss of sensation.

784. autonomic nervous system

A nervous system component that regulates internal organ functions and involuntary processes.

785. axon

The part of a neuron that typically carries impulses away from the cell body toward another neuron.

786. beta blocker

A drug that interferes with sympathetic signals that would ordinarily stimulate beta receptors.

787. biofeedback

Use of signals about levels of autonomic processes to control those processes.

788. catecholamine

A class of amines that act as chemical messengers; dopamine, epinephrine, and norepinephrine.

789. cholinergic

Relating to a neuron whose terminals release acetylcholine.

790. **cholinesterase**

An enzyme that degrades acetylcholine.

791. **cholinesterase inhibitor**

A substance that blocks cholinesterase action.

792. **chromatolysis**

Breakdown of chromatin.

793. **dendrite**

A cytoplasmic process of a neuron that commonly receives signals from other neurons.

794. **depolarization**

Loss of negative charge inside a cell usually associated with the transmission of a nerve impulse.

795. **dopamine**

A neurotransmitter and precursor of norepinephrine.

796. **gamma-aminobutyric acid (GABA)**

An inhibitory neurotransmitter from some neurons of the central nervous system.

797. **habit-forming**

A property that causes some people to make great effort to obtain a drug.

798. **manic-depressive psychosis**

A severe mental disorder characterized by wide mood swings.

799. **membrane potential**

An electrical potential (potential difference) between the inside and outside of a membrane.

800. **multiple sclerosis (MS)**

A central nervous system disease in which tissue hardens and fails to relay impulses.

801. **myelin**

An insulating substance deposited around axons.

802. **myelin sheath**

Insulataing layer that protects most nerves.

803. **myelinated**

Having myelin.

804. **naloxone**

A drug that competes for receptors and counteracts opiate overdoses.

805. **narcosis**

Profound unconsciousness induced by a drug.

806. **neurofibrillary tangles**

Masses of disorderly neural fibers in the brains of Alzheimer patients.

807. **neuropeptide**

A molecule made of a chain of amino acids that influences neural function.

808. **neurotransmitter**

A chemical substance from one neuron that transmits a signal to another neuron at a synapse.

809. **noradrenalin**

Norepinephrine.

810. **norepinephrine**

A neurotransmitter of the sympathetic division of the autonomic nervous system and of some brain neurons.

811. **parasympathetic division**

Autonomic component that accelerates digestion and other functions not essential to a response to stress.

812.   parasympatholytic

Concerning substances that block or counteract sympathetic signals.

813.   parasympathomimetic

Concerning substances that mimic parasympathetic nervous system action.

814.   Parkinson's disease

Muscle rigidity and tremors because of dopamine deficiency in the brain.

815.   physiological dependence

The need for a drug to prevent withdrawal symptoms.

816.   polarized

State of resting membrane that allows it to respond to a stimulus.

817.   postganglionic

Concerning a neuron that receives signals across a synapse in a ganglion; second neuron in an autonomic pathway.

818.   postsynaptic

Referring to a neuron that receives a neurotransmitter at a synapse.

819.   preganglionic

Concerning a neuron that sends signals across a synapse in a ganglion; first neuron in an autonomic pathway.

820.   presynaptic

Referring to a neuron that releases a neurotransmitter at a synapse.

821.   putative

Presumed likely.

822.   schizophrenia

A mental disorder involving a complex set of disturbances in thinking and feeling.

823. **seasonal affective disorder (SAD)**

Depression attributed to lessened light in fall and winter.

824. **serotonin**

A substance secreted as a brain neurotransmitter and a gut hormone.

825. **substantia nigra**

A nucleus of pigmented cells; one of the basal nuclei.

826. **sympathetic division**

The part of the autonomic nervous system that can respond to stressful situations.

827. **sympatholytic**

Blocking or counteracting sympathetic signals.

828. **sympathomimetic**

Mimicking the action of the sympathetic division, or a drug that does so.

829. **tardive dyskinesia**

Involuntary movements probably caused by long-term treatment with tranquilizers.

830. **tolerance**

A condition in which a need for larger doses of a drug develops to get the same effect.

Muscle

831. **actin**

A contractile protein.

832. **action potential**

Wave of change in electrical potential across the cell membrane of an excited cell; impulse.

833. **aldolase**

Simple sugar having its carbonyl carbon at one end of the carbon chain.

834. **anomer**

Stereoisomer of a sugar that differs only in the configuration about the carbonyl carbon atom.

835. **contractile protein**

A protein that acts in shortening a muscle or causing it to develop tension.

836. **contraction cycle**

Repetitive sliding actions of actin and myosin in a muscle filament as it develops tension.

837. **Cori cycle**

Metabolic pathway in which lactic acid moves from muscles to the liver and glucose moves from the liver to muscles.

838. **creatine phosphate**

A molecule that accounts for limited energy storage in muscle.

839. **creatinine**

A metabolic product of creatine excreted at a constant rate in urine.

840. **cross-bridge**

The specialized end of a myosin filament that binds to actin during muscle contraction.

841. **muscular dystrophy**

A disease in which muscles progressively degenerate.

842. **myasthenia gravis**

Progressive muscle weakening because of an autoimmune reaction at motor end plates.

843. **myofibril**

A bundle of thick and thin filaments.

844. **myoglobin**

A pigmented protein that binds oxygen in muscle tissue.

845. **myokinase**

An enzyme that makes ATP and AMP from two molecules of ADP in muscle tissue.

846. **myosin**

A protein that comprises thick filaments of a myofibril.

847. **oxygen debt**

The quantity of oxygen required to oxidize metabolites produced anaerobically during strenuous activity.

848. **phosphocreatine**

An energy storage molecule found in muscle.

849. **rigor mortis**

Muscle rigidity of stiffening following death.

850. **sarcomere**

The contractile unit of skeletal muscle.

851. **sliding filament theory**

An explanation of how myofilaments move with respect to each other during muscle contraction.

852.   tetanus

A sustained contraction maintained by repeated muscle stimulation.

853.   transverse (T) tubule

Crosswise tubule in skeletal muscle myofibrils that carries signals from the sarcolemma to the myofibrils.

854.   tropomyosin

A muscle protein that alters the actin configuration so that contraction can occur.

855.   troponin

A muscle protein that binds to tropomyosin causing it to alter the configuration of actin.

Chemical Senses / Vision

856. **arrestin**

A protein that blocks binding of opsin to transducin in the dark.

857. **chemoreceptor**

A receptor that responds to certain chemical substances.

858. **chlorolabe**

A cone pigment that absorbs green light.

859. **cone**

A light receptor that responds to a certain color.

860. **cyanolabe**

A cone pigment that responds to blue light.

861. **erythrolabe**

A cone pigment sensitive to red light.

862. **gustation**

Sense of taste.

863. **gustatory**

Pertaining to the sense of taste.

864. **hair cell**

A sensory receptor that is stimulated by movement of fluid.

865. **lysozyme**

An enzyme in tears that can destroy microbes.

866. **olfaction**

Sense of smell.

867. **olfactory**

Of the sense of smell.

868. **opsin**

A protein that combines with retinine in the retina.

869. **optic**

Of the eye or concerning the properties of light.

870. **photoreceptor**

A receptor that responds to light.

871. **retinene**

A carotenoid pigment that binds to opsin.

872. **rhodopsin**

A light-sensitive protein found in rods of the retina.

873. **rod**

A receptor in the retina that responds to different intensities of light but not to color.

874. **taste bud**

A structure on the tongue containing taste receptor cells.

875. **transducin**

Enzyme involved in the visual process.

Body Defenses

876. agglutinin

An antibody in an agglutination reaction.

877. agglutinogen

An antigen that elicits an agglutination reaction.

878. antibiotic

Organic compound produced by a living organism that is toxic to certain other species.

879. antibody

A protein elicited by an antigen that can react with and inactivate the antigen.

880. antigen

A substance that elicits a response from the immune system.

881. antigen presenting cell

A macrophage or other cell that processes an antigen and presents it to a B or T lymphocyte.

882. antigenic determinant

The part of an antigen molecule that elicits an immunologic response.

883. B lymphocyte

A lymphocyte that produces plasma cells, which in turn produce antibodies.

884. cell-mediated immunity

Disease resistance involving direct destruction of antigenic cells.

885. chemotaxis

The act of a chemical stimuli to attract or repel; a process that causes some leukocytes to migrate to sites of injury.

886. **clonal selection theory**

Explanation of how lymphocytes are sensitized to a certain antigen and how immune tolerance for self arises.

887. **clone**

A group of genetically like cells derived from a single parent cell.

888. **cloning**

Formation of a clone.

889. **complement**

A group of plasma enzymes that catalyze a sequence of reactions against many different kinds of foreign matter.

890. **cytotoxic**

Harmful to cells.

891. **hapten**

A small molecule that acts as an antigenic determinant when bound to a larger molecule.

892. **helper T cell**

T lymphocyte that facilitates action of B lymphocytes in humoral immunity.

893. **heterogeneity**

Diversity.

894. **histocompatibility complex proteins**

Proteins that give rise to antigenic individuality of a person's cells.

895. **human leukocyte antigen (HLA)**

One of a group of antigens that give cells a unique identity and that are used to match organ donors with recipients.

896. **humoral immunity**

Resistance to disease produced by antibodies.

897.   IgA

An immunoglobulin in secretions.

898.   IgD

An immunoglobulin of unknown function.

899.   IgE

An immunoglobulin responsible for allergic responses.

900.   IgG

An immunoglobulin in blood and primarily responsible for resisting infection.

901.   IgM

A multiunit immunoglobulin most abundant early in an immune response.

902.   immune

Disease resistant.

903.   immune response

Action of lymphocytes sensitized to a particular antigen.

904.   immunity

A state of disease resistance.

905.   immunization

Use of a vaccine or other procedure to create immunity

906.   immunoglobulin

A protein that can bind with a foreign substance; antibody that binds with an antigen.

907.   immunology

The study of immunity and immune reactions.

908.   immunosuppression

A procedure used to lessen an immune response.

909.   immunotoxin

An antibody bound to a toxic drug.

910.   inflammation

Localized response to tissue injury, usually involving redness, swelling, increased temperature, and pain.

911.   interferon

A protein released by virally-infected cells that causes adjacent cells to make an antiviral protein.

912.   interleukin

A substance that facilitates or enhances an immune reaction.

913.   killer T cell

Lymphocyte that directly attacks cells with an antigen it recognizes.

914.   kinin

A substance that elicits events in the inflammatory process.

915.   leukocytosis-promoting (LP) factor

A substance that facilitates migration of leukocytes to a site of injury.

916.   lymphokine

A substance that stimulates activity of lymphocytes.

917.   macrophage

A large phagocytic cell in connective tissue.

918.   major histocompatibility complex proteins

Proteins that give antigenic individuality to a person's cells; on leukocytes, HLAs.

919. **mast cell**

A connective tissue cell that by releasing histamine initiates allergic reactions.

920. **memory cell**

Cells that persist in the immune system after sensitization to recognize future encounters with the same antigen.

921. **monoclonal antibody**

An antibody to an antigen made by clone cells from a sensitized parent cell.

922. **perforin**

A cytotoxic protein that destroys cells by perforating membranes.

923. **phagocyte**

Cell that can engulf and destroy debris, foreign particles, and other cells.

924. **phagocytosis**

Engulfment into a vacuole and digestion by a scavenger cell.

925. **plasma cell**

An antibody-producing cell derived from a B lymphocyte.

926. **primary response**

Initial response to an antigen in which sensitization and memory cell production occurs.

927. **properdin pathway**

A sequence of reactions that activates complement.

928. **psychoneuroimmunology**

A scientific field concerned with effects and interactions of psychological, neurological, and immunological factors.

929. **pus**

A product of inflammation consisting of debris from dead leukocytes and microorganisms.

930. **pyrogen**

A substance that causes body temperature to increase.

931. **secondary response**

Rapid response to an antigen previously encountered.

932. **T cell**

T lymphocyte.

933. **T lymphocyte**

A thymus-processed lymphocyte that can differentiate into several kinds of T cells.

Body Defense Disorders

934. **acquired immune deficiency syndrome (AIDS)**

A viral disease that severely impairs immunity.

935. **agammaglobulinemia**

An immunodeficiency due to a lack of B lymphocytes.

936. **allergen**

A substance capable of eliciting an allergic reaction.

937. **allergy**

Unusual sensitivity to a normally harmless concentration of a substance.

938. **anaphylaxis**

A severe allergic reaction to a substance after prior sensitization to it.

939. **antiviral protein**

A protein produced by cells in response to stimulation by interferon.

940. **atopy**

A tendency to develop hypersensitivity and to produce large numbers of IgEs.

941. **autoantibody**

An antibody against some substance normally present in the body.

942. **autoimmune response**

Release of effector T cells or antibodies that attack a person's own tissues.

943. **blocking antibody**

Antibody that binds to allergen and helps to prevent it from causing an allergic response.

944. **DiGeorge syndrome**

An immunodeficiency due to the absence of T lymphocytes.

945. **graft**

A tissue transplanted to a new site in the same organism or from one organism to another.

946. **graft-versus-host disease**

An immune reaction of graft cells that destroys host cells.

947. **hemolytic disease of the newborn**

Antibodies from a previously sensitized Rh-negative mother destroying erythrocytes in an Rh-positive fetus.

948. **histamine**

A derivative of the amino acid histidine released by injured cells that causes vasodilation and bronchial constriction.

949. **host-versus-graft disease**

An immune reaction in which host cells destroy graft cells.

950. **hypersensitivity**

Abnormal reaction to a substance, as occurs in allergy and certain other immune reactions.

951. **hyposensitization**

Reduction in sensitivity to a substance.

952. **immune complex disorder**

Hypersensitivity in which antigen-antibody complex damages tissues.

953. **immunodeficiency**

An absence or lack of a normal immune function.

954. **sensitization**

Rendering a lymphocyte sensitive to a foreign substance.

955. **severe combined immunodeficiency disease**

An absence of immunity due to a lack of both B and T lymphocytes.

956. **transplant rejection**

An immunologic reaction in which host antigens destroy transplanted tissues or organs.

957. **transplantation**

Moving graft tissue to a new site or new host.

Acid-Base Bal. / Electrolytes / Nitrogenous Wastes

958. **acid-base balance**

Maintenance of body fluid pH within a normal range.

959. **acidosis**

Condition due to low blood pH.

960. **alkalosis**

Condition due to high blood pH.

961. **carbonic anhydrase**

Enzyme that catalyzes reaction in which carbon dioxide and water form carbonic anhydrase or its reverse.

962. **chloride shift**

Movement of chloride ions down an electrical gradient toward a region of positive charge.

963. **clearance**

Rate at which the kidneys can remove a substance from the blood.

964. **conjugate acid-base pair**

A proton donor and its salt.

965. **contraction alkalosis**

Abnormal increase in the blood pH caused by a decrease in body fluid volume.

966. **dialysis**

Separation of molecules by allowing smaller ones to pass through a selectively permeable membrane.

967. **dissociation constant (K)**

Measure of the degree to which a compound ionizes into its components.

968. diuresis

Increase in urine volume.

969. diuretic

Agent that causes diuresis.

970. edema

Excess fluid accumulation in the tissues.

971. electrolyte

Substance that ionizes and conducts electricity.

972. excretion

Elimination of a waste product.

973. fluid regulation

Maintenance of body fluid volumes within normal ranges.

974. glomerular filtration rate (GFR)

Rate at which fluid moves from blood in a glomerulus to kidney filtrate.

975. hemodialysis

The removal of substances from the blood by dialysis.

976. Henderson-Hasselbalch equation

Equation that relates pH to concentrations of an acid and its salt.

977. insensible

Imperceptible.

978. juxtaglomerular apparatus

Cells near a glomerulus that help to regulate blood pressure by releasing the enzyme renin when pressure rises.

979. **loop of Henle**

U-shaped segment of a nephron where sodium chloride becomes concentrated in peritubular fluid.

980. **macula densa**

Modified distal convoluted tubule cells associated with the juxtaglomerular apparatus.

981. **metabolic acidosis**

Low blood pH due to a metabolic (nonrespiratory) disorder.

982. **metabolic alkalosis**

High blood pH due to a metabolic (nonrespiratory) disorder.

983. **nephron**

Functional unit of a kidney.

984. **net filtration pressure**

Pressure pushing materials out of a blood vessel.

985. **osmoreceptor**

Receptor that senses changes in osmolarity.

986. **osmotic diuresis**

Urine volume increase because of large numbers of osmotically active particles in the kidney filtrate.

987. **pK**

Negative logarithm of the equilibrium constant.

988. **renal clearance rate**

Rate at which kidneys remove, or clear, a substance from the blood.

989. **renal failure**

Inability of kidneys to remove wastes and adjust composition of plasma.

990. **renal glycosuria**

Presence of glucose in the urine because of a tubule defect that prevents its return to the blood.

991. **renal hypertension**

Elevated blood pressure because of renal artery constriction.

992. **renal threshold**

Maximum concentration of a substance that can be returned to the blood from the kidney filtrate.

993. **renin**

Kidney secretion that activates angiotensinogen to angiotensin I.

994. **renin-angiotensin mechanism**

A mechanism that increases blood pressure and blood volume when either falls below normal.

995. **respiratory acidosis**

Low blood pH due to a respiratory disorder.

996. **respiratory alkalosis**

High blood pH due to a respiratory disorder.

997. **thirst**

Desire for water or other fluid.

998. **thirst center**

Hypothalamic nucleus that responds to changes in the blood osmotic pressure, causing drinking behavior.

999. **ultrafiltrate**

Filtrate formed under high pressure.

1000. **urea**

Main nitrogenous waste in human urine.

1001. **urea cycle**

Metabolic pathway that synthesizes urea.

1002. **water balance**

A state in which water intake and water output are equal.

Basic Genetics / Mutations

1003. **allele**

One of two or more genes for a particular trait found at a given site on a chromosome.

1004. **autosomal**

Concerning paired (nonsex) chromosomes and the genetic information they carry.

1005. **back-mutation**

Mutation that causes a mutated gene to return to its original wild type.

1006. **deletion**

The loss of one or more bases from a DNA strand.

1007. **dizygotic**

Arising from two separate zygotes.

1008. **dominant**

In genetics, a characteristic that appears in the phenotype whenever its allele is present in the genotype.

1009. **frameshift mutation**

A DNA sequence change caused by adding or deleting bases.

1010. **gene**

Functional unit of heredity; a site on a chromosome that transmits a particular hereditary characteristic.

1011. **genetic information**

Hereditary information stored in a sequence of nucleotide bases.

1012. **genetics**

The study of heredity.

1013. **genotype**

Alleles of a single gene or all the genes carried by a particular individual.

1014. **haploid**

Having one of a pair of chromosomes.

1015. **heterozygous**

Having unlike alleles for a trait.

1016. **homozygous**

Having like alleles for a trait.

1017. **intercalating mutagen**

Agent that causes a mutation by inserting itself between two nucleotides causing a frameshift.

1018. **leaky mutant**

Organism in which mutant gene retains some of its normal activity.

1019. **lethal mutation**

Mutation such that product is so defective as to keep organism from surviving.

1020. **monozygotic**

Arising from the same zygote.

1021. **mutagen**

An agent that can alter DNA.

1022. **mutant**

Gene that has undergone mutation or organism having such a gene.

1023. **phenotype**

The appearance of an individual with respect to one or all inherited characteristics.

1024. **point mutation**

A change in a single base in a DNA molecule.

1025. **recessive**

In genetics, a characteristic seen in a phenotype only when genotype has only the recessive allele.

1026. **recombinant DNA**

DNA derived from a combination of genes from two different organisms.

1027. **recombination**

Combining of genetic material from two different organisms.

1028. **silent mutation**

Change in DNA that fails to produce a detectable change in characteristics of the gene product.

1029. **substitution**

Mutation resulting from replacing one base with another in a DNA molecule.

1030. **suppressor**

Mutation that restores a function lost by way of an earlier mutation, but involving a gene at a different locus.

1031. **thymine dimer**

Two joined thymine residues in a DNA molecule induced by ultraviolet light absorption.

1032. **wild type**

Normal allele or organism having it.

1033. **zygote**

Single cell resulting from a union of ovum and sperm; first cell of a new individual.

Genetic Disorders / Cancer

1034. Ames test

Test for carcinogenicity using growth of certain bacteria.

1035. amniocentesis

A technique for obtaining amniotic fluid for genetic analysis.

1036. amnion

Membrane around a fetus that fills with fluid and acts as a shock absorber.

1037. benign

Nonmalignant; favorable for recovery.

1038. cachexia

Wasting, weakness, and weight-loss seen especially in cancer patients.

1039. cancer

Malignant, invasive tumor that grows by uncontrolled cell division.

1040. carcinogen

A cancer-inducing agent.

1041. chorion

Outermost fetal membrane, which is incorporated into the placenta.

1042. chorionic villi

Tufts of fetal blood vessels across which substances are exchanged with maternal blood.

1043. chorionic villus biopsy

Process of getting fetal tissue during development to look for genetic or developmental defects.

1044. chromosomal abnormality

Detrimental change in the DNA configuration in a chromosome.

1045. chrononcology

The use of cell division rhythms to schedule cancer therapy.

1046. chronotherapy

The use of any rhythms to schedule therapy.

1047. congenital

Present at birth.

1048. cystic fibrosis

An inherited disorder in which thick mucus blocks respiratory and pancreatic passageways.

1049. genetic engineering

Use of human-designed procedures to alter genetic information.

1050. genetic screening

Search for genetic defects in fetuses, newborns, and prospective parents.

1051. hemophilia

An inherited inability to produce a blood clotting factor.

1052. Huntington's chorea

A dominant hereditary disorder that causes degeneration of the nervous system.

1053. inborn error

Biochemical disorder due to an inherited enzyme defect.

1054. Klinefelter's syndrome

A condition due to the presence of XXY sex chromosomes.

1055. malignancy

A tendency to become more virulent; a cancerous growth.

1056. malignant

Having attributes of malignancy.

1057. metastasis

The transfer of disease from one organ to another.

1058. nondisjunction

Failure of replicated chromosomes to separate.

1059. phenylketonuria

A genetic defect in phenylalanine metabolism that causes mental retardation if untreated.

1060. remission

The abatement of disease symptoms or the period during which it occurs.

1061. reproductive engineering

Human-designed procedures used to alter the reproductive process.

1062. rubella

A viral infection commonly called German measles.

1063. teratogen

An agent that causes defective embryonic development.

1064. translocation

Transfer of part of a chromosome from its normal location to a location on another chromosome.

1065. trisomy

Having three copies of a chromosome.

1066. tumor necrosis factor

A substance that causes degeneration and death of tumor cells.

1067. Turner's syndrome

A condition due to having a single X chromosome (without another X or Y chromosome).

Biosphere

1068. acid rain

Rain containing acids from atmospheric pollution.

1069. ammonification

Conversion of protein to ammonia by soil bacteria.

1070. biodegradable

Capable of being decomposed by living organisms.

1071. biogeochemical cycle

Continuous passage of a mineral or other material in an ecosystem.

1072. biological magnification

The concentration of substances in tissues as they are passed up the food chain.

1073. biomass

The total mass of all organisms living in a particular location.

1074. biosphere

All living matter on land, in water, and in the atmosphere.

1075. carbon cycle

A repetitive sequence of chemical processes in which carbon enters and leaves living organisms.

1076. denitrification

The process of converting nitrate to nitrogen gas; a part of the nitrogen cycle.

1077. desalinization

Removal of salt, typically from ocean water.

1078. **environmental impact statement**

A statement of the findings of a detailed study of how an activity might affect the environment.

1079. **euryhaline**

Ability to tolerate a variety of salinities in the environment.

1080. **eutrophication**

The process of aging and death of organisms in a pond or lake.

1081. **fossil fuel**

Combustible material, such as coal, oil, and gas, derived from previously living material.

1082. **greenhouse effect**

An increase in environmental temperature as carbon dioxide absorbs heat radiated from the earth.

1083. **groundwater**

Water in soil and in aquifers beneath the soil.

1084. **hazardous waste**

Environmental pollutants that endanger living things.

1085. **hydrothermal vent**

A volcanic opening on the ocean floor from separation of tectonic plates that spews forth hot water and minerals.

1086. **lithosphere**

Mineral portion of the earth's surface.

1087. **mariculture**

Growing of human food in the ocean.

1088. **marine**

Pertaining to the ocean.

1089. nitrification

Formation of nitrates by bacteria.

1090. nitrogen cycle

Continuous passage of various forms of nitrogen through a series of living things.

1091. nitrogen fixation

Conversion of atmospheric nitrogen to biologically useful form by nitrogen-fixing organisms.

1092. nitrogenase system

Enzyme system capable of reducing atmospheric nitrogen to ammonia using energy from ATP.

1093. nuclear waste

Unused radioactive material from nuclear power plants and other processes involving radiation.

1094. osmoconformer

Marine organism whose body takes on the same salt concentration as the environment.

1095. ozone shield

A layer of ozone in the upper atmosphere that protects the earth from damaging ultraviolet radiation.

1096. photoperiodism

Physiological responses to light and darkness.

1097. pollution

The presence of substances that damage the environment.

1098. primary treatment

The first treatment given to sewage.

1099. salinity

Quantity of dissolved salts in a solution (in parts per thousand).

1100. **salinization**

Deposition of salt.

1101. **secondary treatment**

Bacterial decomposition in a sewage treatment plant.

1102. **siltation**

Deposition of silt.

1103. **stenohaline**

Property that allows animals to live in only a restricted range of environmental salt concentrations.

1104. **temperature inversion**

An event that occurs when air near the ground is cooler than air above it.

1105. **tertiary treatment**

A process that removes toxic substances and excess minerals from sewage effluent.

1106. **thermal pollution**

Abnormally high temperature produced in an environment.

1107. **trophallaxis**

Mutual exchange of food and secretions among members of an insect colony.

1108. **water cycle**

A sequence of repetitive reactions in which water moves to and from living things.

1109. **water table**

A level below which aquifers are filled with water.

Index

absolute configuration	7	aminopeptidase	52
absorption	52	aminotransferase	22
absorption spectrum	36	ammonification	112
absorptive	52	amniocentesis	108
accessory pigment	36	amnion	108
acetylcholine	81	amphibolic pathway	30
acid	1	amphipathic compound	7
acid mucopolysaccharide	78	amphoteric compound	7
acid rain	112	amylase	52
acid-base balance	100	anabolic	30
acidosis	100	anabolic steroid	63
acquired immune deficiency		anabolism	30
syndrome (AIDS)	97	anaerobe	30
actin	86	anaerobic	30
actinomycin D	46	anaphylaxis	97
action potential	86	anaplerotic reaction	30
action spectrum	36	androgen	63
activation energy	27	anemia	72
active site	27	anesthetic	81
active transport	18	angiotensin	63
activity	30	angiotensinogen	63
activity coefficient	30	Angstrom unit	7
adenine	22	anion	1
adenosine	22	anomer	86
adenosine diphosphate	22	anorexia	57
adenosine triphosphate	22	anorexia nervosa	57
adipose	78	anoxia	70
adrenal	63	antibiotic	91
adrenalin	63	antibody	91
adrenergic	81	anticoagulant	72
adrenocorticotropic hormone	63	anticodon	46
adsorptive endocytosis	18	antidiuretic hormone (ADH)	63
aerobe	30	antigen	91
aerobic	30	antigen presenting cell	91
agammaglobulinemia	97	antigenic determinant	91
agglutinin	91	antiparallel	41
agglutinogen	91	antithrombin	72
alarm reaction	63	antiviral protein	97
albumin	72	aortic body	70
alcoholism	57	apoferritin	72
aldolase	86	argentaffin cells	52
aldosterone	63	arrestin	89
alkaline	1	ascorbic acid	57
alkaloid	7	asthma	70
alkalosis	100	asymmetric carbon atom	7
allele	105	atelectasis	70
allergen	97	atherosclerosis	72
allergy	97	atom	1
allosteric enzyme	27	atomic number	1
allosteric site	27	atomic symbol	1
Alzheimer's disease	81	atomic weight	1
Ames test	108	atopy	97
amino acid	7	ATP synthetase	31
amino acid activation	46	ATPase	31
aminoacyl-tRNA	46	atrial natriuretic hormone (ANH)	64
aminoacyl-tRNA synthetase	46	autoantibody	97

autoimmune response	97	capsid	41
autonomic nervous system	81	carbaminohemoglobin	70
autosomal	105	carbohydrate	7
autotroph	31	carbohydrate-craving obesity	58
auxotroph	31	carbon cycle	112
Avogadro's number	1	carbonic anhydrase	100
axon	81	carboxypeptidase	52
B lymphocyte	91	carcinogen	108
B-DNA	41	cardiopulmonary	
back-mutation	105	resuscitation (CPR)	73
bacteriochlorophyll	36	cardiovascular	73
bacteriophage	41	carotene	58
basal metabolic rate (BMR)	57	carotenoid	36
basal metabolism	57	carrier	18
base	1	carrier saturation	18
base pair	41	cartilage	78
base pairing	46	catabolic	31
benign	108	catabolism	31
beta blocker	81	catabolite activator protein	46
beta conformation	22	catalyst	2
beta lipotropin	64	catalytic site	27
beta oxidation	22	catecholamine	81
beta reduction	22	cation	2
bilayer	22	CCK (cholecystokinin-	
bile	52	pancreozymin)	52
bile salt	52	cell	13
bilirubin	72	cell membrane	13
biliverdin	72	cell theory	13
binding site	18	cell-mediated immunity	91
biodegradable	112	cellular respiration	31
bioenergetics	31	cellulose	8
biofeedback	81	central dogma	41
biogeochemical cycle	112	centriole	13
biological magnification	112	cerebrospinal fluid	73
biomass	112	ceruloplasmin	73
biomolecule	7	CF-one	36
biosphere	112	CF-zero	37
biotin	57	Chargaff's rules	41
blocking antibody	97	chemical bond	2
blood	72	chemical energy	2
blood-brain barrier	72	chemiosmotic coupling	31
Bohr effect	70	chemiosmotic theory	31
bond energy	31	chemoreceptor	89
Boyle's law	70	chemotaxis	91
bradykinin	73	chenodeoxycholic acid	53
brown fat	52	chief cell	53
buffer	1	chimeric DNA	41
building-block molecule	7	chimeric protein	46
bulimia	57	chloride shift	100
bulk flow	18	chlorolabe	89
bundle sheath cell	36	chlorophyll	37
C-four plant	36	chloroplast	37
C-three plant	36	cholesterol	8
cachexia	108	cholesteryl esterase	22
calciferol	57	cholic acid	53
calcitonin	64	cholinergic	81
calmodulin	64	cholinesterase	82
calorie	57	cholinesterase inhibitor	82
Calvin cycle	36	chorion	108
cancer	108	chorionic villi	108

chorionic villus biopsy	108	creatine phosphate	86
chromatin	13	creatinine	86
chromatolysis	82	cross-bridge	86
chromosomal abnormality	109	cross-matching	73
chromosome	13	crossover point	32
chrononcology	109	cyanocobalamin	58
chronotherapy	109	cyanolabe	89
chylomicron	53	cyclic AMP	64
chyme	53	cyclic electron flow	37
chymotrypsin	53	cyclic photophosphorylation	37
chymotrypsinogen	53	cystic fibrosis	109
cilium	13	cytochrome	32
cisterna	13	cytoplasm	13
citric acid cycle	23	cytosine	23
clearance	100	cytoskeleton	13
clonal selection theory	92	cytosol	14
clone	92	cytotoxic	92
cloning	92	Dalton	2
clot retraction	73	Dalton's law	70
co-transport	53	dark reaction	37
codon	46	deamination	23
coenzyme	27	degenerate code	47
coenzyme A	58	dehydration	8
cofactor	27	dehydrogenase	27
cohesive end	46	deletion	105
colinearity	47	denaturation	23
colipase	53	denatured protein	23
collagen	18	dendrite	82
colligative property	2	denitrification	112
colloid	2	deoxyribonuclease	42
colloidal dispersion	2	deoxyribonucleic acid (DNA)	42
colloidal osmotic pressure	18	deoxyribose	23
common intermediate	32	depolarization	82
competitive inhibition	27	depurination	47
complement	92	desalinization	112
complementary base pairing	41	dextrorotatory isomer	8
complementary DNA	47	diabetes insipidus	64
compound	2	diabetes mellitus	58
concentration gradient	18	dialysis	100
cone	89	diasteriosiomer	8
configuration	8	diffuse endocrine system (DES)	64
conformation	8	diffusion	18
congenital	109	DiGeorge syndrome	98
conjugate acid-base pair	100	digestion	53
conjugated protein	23	dipeptidase	53
consensus sequence	47	dipeptide	54
constitutive enzyme	27	dipole	2
constitutive heterochromatin	41	diprotic acid	3
contractile protein	86	disaccharide	8
contraction alkalosis	100	dissociation constant (K)	100
contraction cycle	86	disulfide bridge	28
coordinate induction	47	diuresis	101
core body temperature	58	diuretic	101
core particle	47	dizygotic	105
Cori cycle	86	DNA chimera	42
corticosterone	64	DNA glycosidase	42
cortisol	64	DNA ligase	42
coumarin	73	DNA polymerase	47
coupled reaction	32	DNA replicase system	42
covalent bond	2	DNA replication	42

dominant	105	exon	48
dopamine	82	exonuclease	42
double helix	42	extracellular	19
dynorphin	64	facilitated diffusion	19
edema	101	facultative	33
elastin	78	facultative heterochromatin	48
electrochemical gradient	19	fatty acid	8
electrolyte	101	fermentation	33
electron	3	ferritin	74
electron acceptor	32	fibrin	74
electron carrier	32	fibrinogen	74
electron donor	32	fibroblast	78
electron transport system	32	fibrocyte	19
eleidin	78	filtration	19
element	3	flagellum	14
elongation factor	47	flavin adenine dinucleotide (FAD)	33
Emerson enhancement effect	37	fluid pinocytosis	19
emphysema	70	fluid regulation	101
enantiomer	8	fluid-mosaic model	19
end-product inhibition	47	fluorescence	37
endergonic	32	folacin	58
endocrine	64	follicle-stimulating hormone (FSH)	65
endocytosis	19	fossil fuel	113
endolymph	73	frameshift mutation	105
endonuclease	42	functional group	9
endoplasmic reticulum	14	gamma-aminobutyric acid (GABA)	82
endorphin	65	gastrin	65
energy charge	32	gastroferrin	74
energy coupling	32	gel	14
enkephalin	65	gene	105
enteric	65	general adaptation syndrome	65
enterokinase	54	genetic code	48
enthalpy	33	genetic engineering	109
entropy	33	genetic information	105
environmental impact statement	113	genetic screening	109
enzyme	28	genetics	105
enzyme repression	47	genome	43
epimer	8	genotype	106
epimerase	28	gliadin	54
epinephrine	65	globin	74
equilibrium	33	globulin	74
equilibrium constant	33	glomerular filtration rate (GFR)	101
erythrocyte	73	glomerulus	74
erythrolabe	89	glucagon	65
erythropoiesis	73	glucocorticoid	66
erythropoietin	74	gluconeogenesis	33
essential amino acid	58	glucose	9
essential fatty acid	58	glucose sparing	33
estradiol	65	gluten	54
estrogen	58	glycerol	9
euchromatin	42	glycine	23
eukaryote	14	glycogenesis	33
euryhaline	113	glycogenolysis	34
eustress	65	glycolate pathway	37
eutrophication	113	glycolipid	9
excited state	37	glycolysis	34
excretion	101	glycoprotein	19
exergonic	33	goblet cell	78
exhaustion stage	65	Golgi apparatus	14
exocytosis	19	gonadotropin	66

gout	48	hydrolysis	9
gradient	19	hydronium ion	3
graft	98	hydrophilic	14
graft-versus-host disease	98	hydrophobic	14
gram molecular weight	3	hydrophobic interaction	14
granum	37	hydrostatic pressure	20
greenhouse effect	113	hydrothermal vent	113
ground state	38	hydroxyapatite	78
ground substance	78	hyperglycemia	59
groundwater	113	hyperosmotic	20
growth hormone	66	hypersensitivity	98
growth hormone-hypothalamic		hypertension	71
mechanism	58	hyperthermia	59
guanidine triphosphate (GTP)	34	hypertonic	20
guanine	48	hypoglycemia	59
gustation	89	hypoglycemic	59
gustatory	89	hypophysis	66
habit-forming	82	hyposensitization	98
hair cell	89	hyposmotic	20
haploid	106	hypothermia	59
hapten	92	hypotonic	20
hazardous waste	113	hypoxia	71
helix	43	IgA	93
helper T cell	92	IgD	93
hematocrit	74	IgE	93
hematopoiesis	74	IgG	93
heme	70	IgM	93
hemodialysis	101	immune	93
hemoglobin	71	immune complex disorder	98
hemolysis	74	immune response	93
hemolytic	75	immunity	93
hemolytic disease of the newborn	98	immunization	93
hemophilia	109	immunodeficiency	98
hemopoietic	75	immunoglobulin	93
hemorrhage	75	immunology	93
hemosiderin	75	immunosuppression	94
hemostasis	75	immunotoxin	94
Henderson-Hasselbalch equation	101	in vitro	14
heparin	75	in vivo	14
hepatitis	54	inborn error	109
heterogeneity	92	induced enzyme	48
heterozygous	106	induced fit	28
Hill reaction	38	inducer	48
hirudin	75	inflammation	94
histamine	98	informational molecule	9
histocompatibility complex proteins	92	initiation codon	48
histone	23	initiation complex	48
homologous protein	23	initiation factor	48
homotropic enzyme	28	insensible	101
homozygous	106	insertion	43
hormone	66	insertion sequence	43
host-versus-graft disease	98	insertional mutagenesis	43
human chorionic gonadotropin	66	insulin	66
human leukocyte antigen (HLA)	92	insulin shock	59
humoral immunity	92	insulin-glucagon mechanism	59
Huntington's chorea	109	integral	23
hyaline membrane disease (HMD)	71	intercalating mutagen	106
hydrocephalus	75	interferon	94
hydrochloric acid	54	interleukin	94
hydrogen bond	9	intermediary metabolism	34

intermediate	34	macromolecule	10
intracellular	15	macrophage	94
intrathylakoid space	38	macula densa	102
intrinsic factor	54	major groove	43
intron	48	major histocompatibility complex	
ion	3	proteins	94
ionic bond	3	malignancy	110
ionizing radiation	3	malignant	110
irreversible process	34	malnutrition	59
ischemia	75	maltase	54
islet of Langerhans	66	manic-depressive psychosis	82
isoelectric pH	24	marasmus	60
isomer	9	mariculture	113
isomerase	28	marine	113
isoprene	24	mast cell	95
isosmotic	20	melanin	79
isothermal process	34	membrane potential	82
isotonic	20	memory cell	95
isotope	3	mesophyll cell	38
isozyme	28	messenger RNA (mRNA)	49
jaundice	75	metabolic acidosis	102
juxtaglomerular apparatus	101	metabolic alkalosis	102
keratin	78	metabolic rate	60
keratohyalin	79	metabolic turnover	15
ketogenic amino acid	24	metabolic water	24
ketone body	24	metabolism	34
ketose	24	metabolite	34
ketosis	24	metalloenzyme	28
killer T cell	94	metastasis	110
kilocalorie	59	micelle	55
kinase	28	Michaelis constant (Km)	28
kinetic	34	Michaelis-Menten equation	29
kinin	94	microfilament	15
Klinefelter's syndrome	109	microsome	15
kwashiorkor	59	microtubule	15
lactase	54	microvillus	55
lacteal	54	mineral	60
lactic acid	9	mineralocorticoid	66
law of mass action	3	minor groove	43
leaky mutant	106	mitochondrion	15
lecithin	9	mixture	3
lethal mutation	106	modulator	49
leukocytosis-promoting (LP) factor	94	molal solution	4
levorotatory isomer	9	molar solution	4
ligand	20	mole	4
light reaction	38	molecule	4
Lineweaver-Burk equation	28	monoclonal antibody	95
lipase	54	monolayer	24
lipid	10	monoprotic acid	4
lipoic acid	59	monosaccharide	10
lipoprotein	24	monozygotic	106
lithosphere	113	mucin	79
loop of Henle	102	mucopolysaccharide	79
low-energy phosphate	34	mucoprotein	79
lutein	66	mucous	79
luteinizing hormone (LH)	66	mucus	79
lymph	75	multienzyme system	29
lymphokine	94	multiple sclerosis (MS)	82
lysosome	15	muscular dystrophy	87
lysozyme	89	mutagen	106

mutant	106	nucleus	4
mutarotation	24	nutrition	60
mutase	29	obesity	60
mutation	43	olfaction	90
myasthenia gravis	87	olfactory	90
myelin	83	oligomeric protein	25
myelin sheath	83	oligosaccharide	25
myelinated	83	open system	35
myofibril	87	operator	49
myoglobin	87	operon	49
myokinase	87	opsin	90
myosin	87	optic	90
naloxone	83	optical activity	10
narcosis	83	optimum pH	29
native conformation	10	organelle	15
natriuresis	67	organic	4
negative supercoil	43	orthophosphate cleavage	35
nephron	102	osmoconformer	114
net filtration pressure	102	osmolarity	20
net protein utilization	60	osmoreceptor	102
neurofibrillary tangles	83	osmosis	16
neurohypophysis	67	osmotic diuresis	102
neuropeptide	83	osmotic pressure	20
neurotransmitter	83	osteoporosis	79
neutron	4	oxidation	4
niacin	60	oxidation-reduction	4
nicotinamide adenine dinucleotide		oxidative phosphorylation	35
(NAD)	35	oxidizing agent	4
nicotinamide adenine dinucleotide		oxygen debt	87
phosphate	38	oxygenase	29
nitrification	114	oxygenic photoautotroph	38
nitrogen balance	49	oxyhemoglobin	71
nitrogen cycle	114	oxytocin	67
nitrogen fixation	114	ozone shield	114
nitrogenase system	114	P/O ratio	38
noncompetitive inhibition	29	packing ratio	44
noncyclic electron flow	38	palindromic	44
noncyclic photophosphorylation	38	pancreas	55
nondisjunction	110	pancreatic	55
nonessential amino acid	24	pancreatic juice	55
nonheme-iron protein	49	pancreatitis	55
nonhistone chromosomal protein	43	pantothenic acid	60
nonpolar	10	parasympathetic division	83
nonpolar group	10	parasympatholytic	84
nonreiterated sequence	43	parasympathomimetic	84
nonsense codon	49	parathormone	67
nonsense mutation	44	parathyroid glands	67
noradrenalin	83	Parkinson's disease	84
norepinephrine	83	partial pressure	71
nuclear	15	passive transport	20
nuclear waste	114	pentose	25
nuclease	44	pentose phosphate pathway	35
nucleic acid	10	pepsin	55
nucleolus	15	peptidase	25
nucleoplasm	15	peptide bond	25
nucleoside	25	perfluorocarbon	71
nucleoside diphosphate sugar	44	perforin	95
nucleoside diphosphokinase	44	perfusion	76
nucleosome	44	permeability	21
nucleotide	25	pernicious anemia	76

peroxisome	16	postabsorptive	61
pH	5	postganglionic	84
phagocyte	95	postsynaptic	84
phagocytosis	95	posttranslational modification	49
phenotype	106	potential energy	35
phenylketonuria	110	preganglionic	84
phosphocreatine	87	presynaptic	84
phosphogluconate pathway	35	primary response	95
phospholipid	10	primary structure of a protein	25
phosphorylation	35	primary treatment	114
photochemical reduction	38	procarboxypeptidase	55
photoexcitation	39	product	5
photolysis	39	progesterone	67
photon	39	prokaryote	16
photoperiodism	114	prolactin	67
photophosphorylation	39	properdin pathway	95
photoreceptor	90	prostaglandin	67
photoreduction	39	protein	11
photorespiration	39	prothrombin	76
photosynthesis	39	proton	5
photosynthetic phosphorylation	39	protoplasm	16
photosynthetic unit	39	psychoneuroimmunology	95
photosystem	39	puberty	67
photosystem I	39	purine	26
photosystem II	40	puromycin	49
phycobilin	40	pus	95
physiological dependence	84	putative	84
pica	60	pyrimidine	26
pigment	40	pyrogen	96
pinocytosis	16	quantum	40
pituitary gland	67	quantum requirement	40
pK	102	radiation	5
plasma	76	radioactivity	5
plasma cell	95	radioisotope	5
plasma membrane	21	reactant	5
plasma protein	76	reaction center	40
plasmid	44	receptor	21
plasmin	76	recessive	107
plasminogen	76	recombinant DNA	107
plastid	16	recombination	107
plastocyanin	40	reduction	5
plastoquinone	40	reiterated sequence	44
platelet	76	relaxed state	45
pleated sheet	25	relaxin	67
point mutation	107	releasing factor	50
polar compound	10	remission	110
polar group	5	renal clearance rate	102
polarity	44	renal failure	102
polarized	84	renal glycosuria	103
pollution	114	renal hypertension	103
polydipsia	60	renal threshold	103
polymer	10	renin	103
polynucleotide	49	renin-angiotensin mechanism	103
polypeptide	25	repair endonuclease	45
polyphagia	60	repair synthesis	45
polyribosome	49	replication	45
polysaccharide	11	replicon	45
polyuria	61	reproductive engineering	110
porphyrin	25	resistance stage	68
positive supercoil	44	respiration	71

respiratory acidosis	103	steroid	11
respiratory alkalosis	103	stoma	40
respiratory distress syndrome	71	streptokinase	77
respiratory quotient (RQ)	61	stress	68
restriction enzyme	45	stressor	68
restriction site	45	structural gene	50
retinene	90	substantia nigra	85
rhodopsin	90	substitution	107
riboflavin	61	sucrase	56
ribonuclease	55	sudoriferous	80
ribonucleic acid (RNA)	50	sudoriferous gland	80
ribosomal RNA (rRNA)	16	supercoil	45
ribosome	16	suppressor	107
rickets	79	surface tension	21
rigor mortis	87	surface-to-volume ratio	16
rod	90	surfactant	71
rubella	110	suspension	5
salinity	114	sympathetic division	85
salinization	115	sympatholytic	85
salt	5	sympathomimetic	85
sarcomere	87	T cell	96
saturated fatty acid	11	T lymphocyte	96
saturation	11	tardive dyskinesia	85
schizophrenia	84	target cell	68
scurvy	61	taste bud	90
seasonal affective disorder (SAD)	85	taurine	56
sebaceous	79	temperature inversion	115
sebum	79	template	11
second messenger	68	teratogen	110
secondary response	96	termination codon	50
secondary treatment	115	termination sequence	50
secretagogue	55	tertiary treatment	115
secretin	55	testosterone	68
secretion	11	tetanus	88
selectively permeable	21	thalassemia	77
selenosis	61	thermal pollution	115
sensitization	98	thiamine	61
sequence homology	50	thirst	103
serotonin	85	thirst center	103
serum	76	thrombin	77
severe combined immunodeficiency		thylakoid	40
disease	99	thymine	26
sickle cell anemia	76	thymine dimer	107
silent mutation	107	thymosin	68
siltation	115	thymus gland	68
sliding filament theory	87	thyroid gland	68
sodium-potassium pump	16	thyroid-stimulating hormone (TSH)	69
sol	16	tissue plasminogen activator (tPA)	77
solute	21	tissue thromboplastin	77
solution	21	tocopheral	61
solvent	21	tolerance	85
somatostatin	68	tonicity	21
somatotropin	68	topoisomerase	45
spacer sequence	45	total parental nutrition (TPN)	61
specific heat	35	trace element	11
specificity	11	transamination	50
spectrin	76	transcription	50
sprue	56	transcriptional control	50
stenohaline	115	transducin	90
stereoisomer	11	transfer RNA (tRNA)	50

transferrin	77
translation	50
translational control	51
translocation	110
transplant rejection	99
transplantation	99
transverse (T) tubule	88
triacylglycerol	11
triglyceride	12
trisomy	110
trophallaxis	115
tropic	69
tropomyosin	88
troponin	88
trypsin	56
trypsinogen	56
tubulin	17
tumor	17
tumor necrosis factor	111
Turner's syndrome	111
turnover	29
ulcer	56
ulcerative	56
ultrafiltrate	103
unsaturated fatty acid	12
uracil	26
urea	103
urea cycle	104
uridine triphosphate (UTP)	35
valence	6
vasopressin	69
vesicle	17
villikinin	56
villus	56
vitamin A	61
vitamin D	61
vitamin E	62
vitamin K	62
water balance	104
water cycle	115
water table	115
wild type	107
wobble	51
xanthine	51
zygote	107

www.ingramcontent.com/pod-product-compliance
Lightning Source LLC
Chambersburg PA
CBHW081131170526
45165CB00008B/2629